IMPROVING TECHNIQUES IN TEACHING ENGLISH FOR THE JOB

IMPROVING TECHNIQUES IN TEACHING ENGLISH FOR THE JOB

Ross Macdonald
Rudolph Troike
Mary Galvan
Alexa McCray
Len Shaefer
Paul Stupp

Prepared by:

InterAmerica Research Associates, Inc.
1555 Wilson Boulevard
Rosslyn, Virginia 22209

May 1982

DISCRIMINATION PROHIBITED -- Title VI of the Civil Rights Act of 1964 states: "No person in the United States shall, on the grounds of race, color, or national origin, be denied the benefits of, or be subjected to, discrimination under any program or activity receiving financial assistance."

Therefore, each vocational education program, like all other programs or activities receiving financial assistance from the Department of Education, must be operated in compliance with this law.

The presentation given herein was performed pursuant to Contract number 300-80-0810 of the United States Department of Education. However, the opinions presented herein do not necessarily reflect the position or policy of the United States Department of Education, and no official endorsement by the United States Department of Education should be inferred.

This document is published by InterAmerica Research Associates, Inc., pursuant to contract NIE 400-80-0040 to operate the National Clearinghouse for Bilingual Education. The National Clearinghouse for Bilingual Education is jointly funded by the National Institute of Education and the Office of Bilingual Education and Minority Languages Affairs, U.S. Department of Education. Contractors undertaking such projects under government sponsorship are encouraged to express their judgment freely in professional and technical matters; the views expressed in this publication do not necessarily reflect the views of the sponsoring agencies.

InterAmerica Research Associates, Inc. d/b/a
National Clearinghouse for Bilingual Education
1300 Wilson Boulevard, Suite B2-11
Rosslyn, Virginia 22209
(703) 522-0710 / (800) 336-4560

ISBN: 0-89763-069-6
First printing 1982
Printed in USA

10 9 8 7 6 5 4 3 2 1

ACKNOWLEDGMENTS

The authors wish to express their sincerest appreciation to Dr. Richard Naber, Branch Chief for Postsecondary and Vocational Program Development of the Office of Bilingual Education and Minority Languages Affairs, U.S. Department of Education, and to Ms. Joan Cassidy, Program Manager for Bilingual Vocational Training, who shared responsibility with Dr. Naber during the course of the project. We appreciate their interest and support, without which the results presented herein could not have come to fruition.

Special thanks are also due to Dr. Doris Gunderson, Office of Occupational and Adult Education, whose early interest in the work reflected here led to its development; to Ms. Nancy Boss of Miami-Dade Community College, whose input contributed to Appendix B; to Ms. Annette Zarian of UCLA-Extension; and to Mr. Dennis Altfest of Bronx Community College and Mr. Manuel Hernandez of Solidaridad Humana in New York City, whose field-testing of the first draft of this handbook contributed a great deal to making the present version more usable. We are also grateful to Mr. Rafael Martinez and Ms. Sarah Bingaman of Nuevos Horizontes in Arlington Heights, Illinois, Dr. Dorothy Messerschmitt of the University of San Francisco, Dr. Michael Tang of Metropolitan State College in Denver, and Dr. Richard Todd of New York University who provided space for the workshops which were held to disseminate information and obtain comments on the handbook draft; and to the participants in the workshops whose insightful observations guided many of the revisions which were made. Thanks are due to Dr. Richard Hsu of the China Institute in America in New York City for making the facilities of the Institute available for the final meeting of the Advisory Panel, and to Mr. Stuart McKenty for his presentations in several of the information workshops.

We particularly want to thank members of the Advisory Panel, who contributed valuable suggestions and insights from their respective fields of expertise. The Advisory Panel consisted of: Dr. Rupert Evans, University of Illinois; Dr. Barbara Flores, Arizona State University; Mr. Richard Kennedy, Pan American Health Organization; Mr. Jeffrey Kiely, formerly of the Ramah Navajo Vocational Training Program; Dr. Richard Hsu, China Institute in America; Dr. Ramon Santiago, National Association for Bilingual Education; and Dr. John Tsu, University of San Francisco.

Special recognition goes to Ms. Jane Itzel and Ms. Cynthia Swiggard of InterAmerica for their patience and good humor in the preparation of this manuscript, to Ms. Susan Matson for her editorial contribution, and to the Georgetown University Computer Center, without which the project would not have been possible. Finally, we owe particular appreciation to Mr. Ismael Lugo, who successfully saw the project through its first year as coordinator before becoming Associate Dean for Continuing Education at the InterAmerican University of Puerto Rico.

CONTENTS

PREFACE

Using a non-technical presentation, the authors of Improving Techniques in Teaching English for the Job show how computer analysis of job-relevant text materials can be used in developing language instruction for limited-English-proficient students. This practical approach is also designed to help the ESL instructor coordinate activities with the vocational curriculum. In addition, the special applications of data processing to functional language teaching -- procedures for obtaining and using various types of text analysis -- are appropriate to developing materials for native speakers of English whose language skills may need to be strengthened.

One of the activities of the National Clearinghouse for Bilingual Education is to publish documents addressing the specific information needs of the bilingual education community. We are pleased to add this distinguished title to our growing list of publications. In subsequent Clearinghouse products we will similarly seek to contribute information and knowledge that can assist in the education of minority language and culture groups in the United States.

NATIONAL CLEARINGHOUSE
FOR BILINGUAL EDUCATION

FOREWORD TO THE ESL/ENGLISH INSTRUCTOR

As an instructor, you will find that most of this handbook is directed particularly to you, and its results will affect your work more than that of anyone else. The handbook is intended basically to show you how computer outputs can simplify your work, enable you to spend your time more effectively, and especially, permit you to integrate your teaching more closely with the vocational component of the training program. Research on second language learning and prior experience in Bilingual Vocational Training (BVT) programs suggest that adults learn language best when it is closely related to content, rather than taught as an end in itself. Functional language teaching is very demanding, but the computer can help, as you will see.

No prior knowledge of computers is required to follow the discussion presented here. However, since the application of computer processing given here is relatively new, even those familiar with computers should find it of interest. Only a minimal amount of computer terminology is introduced, and this is carefully explained. The emphasis is on the *products*--analyzed texts--of computer processing, and on how to apply these products in teaching job-related English, not on understanding computer operations. This is not to say that no such knowledge is desirable, since the more you know the better you can communicate with computer programmers and others who are not familiar with *your* needs and problems. (This is a striking example of the communication problems even native English speakers face when they do not share a common body of knowledge--language by itself is clearly not enough.)

If you are unfamiliar with bilingual vocational training, you will find it helpful to read Appendix B. The last two sections of Chapter IV ("Cost Considerations in Creating Text Analyses" and "Obtaining Existing Text Analyses") are intended primarily for administrators or computer specialists, and are not essential to follow in detail, although a general understanding is desirable and would promote better communication with those responsible for obtaining computer services for you. Depending on the extent to which you deal with grammatical issues in your teaching, you may find it helpful to follow the discussion of grammatical codes in Appendix A.

As a number of people have pointed out, the principles discussed here apply equally well to teaching native English speakers whose language skills are limited and need developing. The potential scope of interest is thus much larger than just those teaching speakers of other languages.

It is our sincere hope that you will find this handbook helpful, and that it contributes to your efforts to teach job-required English language skills more effectively to persons of limited English-speaking ability who need them for access to employment in our society.

FOREWORD TO THE ADMINISTRATOR

Although this handbook is addressed primarily to job-related ESL instructors in Bilingual Vocational Training programs, there is a great deal here that is of relevance to you, particularly at the stage of planning or budgeting a program. If you are not generally familiar with BVT programs, you will find useful background information in Chapter I and Appendix B, which should also help give you an overview of the role of ESL in such a program.

In anticipation that you may be asked by the job-related ESL instructor to budget funds for computer processing, or to arrange for computer services, you should read sections B and C of Chapter IV ("Cost Considerations in Creating Text Analyses" and "Obtaining Existing Text Analyses"), and examine Appendix C carefully. However, in order to better understand exactly what you may be asked to do, and to be able to evaluate the request, you would find it most helpful to read Chapter II and section A of Chapter IV.

As an administrator, you are of course concerned about the effectiveness of your program. The procedures discussed in this handbook can make the instructors in your program more effective in developing students' language skills; for this reason you may wish to bring the handbook to their attention if they are not already aware of it. You may decide to plan in your budget to include the necessary funds for computer processing, and you may well want to arrange for inservice staff development to prepare them to utilize the computer analyses discussed here.

It is our hope that you will find this handbook helpful, and that it contributes to making your program more effective in preparing trainees for the job. It is written as non-technically as possible, so that no prior acquaintance with computers is necessary to follow the discussion. Should you have additional questions, you should write or call the Office of Bilingual Education and Minority Languages Affairs in the Department of Education, or the National Clearinghouse for Bilingual Education (800-336-4560 toll free), for further information.

FOREWORD TO THE VOCATIONAL INSTRUCTOR

When the project on which this handbook is based was started, it was thought that the topic it deals with--using the computer in analyzing technical text material as a basis for developing more relevant language instruction for limited English speakers--was of interest only to job-related ESL instructors. As the project progressed, however, more and more vocational instructors told us that the procedures discussed here were relevant to vocational instruction itself in a variety of situations. Thus you may find that there is more of interest here than meets the eye, in terms of the issues directly addressed. We hope that this may be the case, and encourage you to think imaginatively about what you find in the handbook.

More immediately, the thrust of the handbook is to help the job-related ESL instructor bring the curriculum of the language teaching component into closer coordination with the content of the vocational curriculum. Since the language instructor rarely has expertise in the vocational area, it will be necessary for him or her to work closely with you and to seek your input into the content of the language lessons. Inasmuch as you may lack training in the language field, this handbook can give you a better idea of the needs and concerns of the job-related ESL instructor, and help you in working out a more successful means of cooperating in your joint endeavor. Since language teaching and vocational instruction should mutually reinforce one another in a bilingual vocational training program, your students will stand to gain a great deal from a more closely integrated program.

It will be most immediately useful for you to read Chapter I, "Introduction and Background", and Appendix B, "English for the Job in the Bilingual Vocational Training Program." If you then review Chapter II, you will be better prepared to respond to specific questions the ESL instructor might bring to you about the material you are teaching, including helping in the selection of material to be submitted for computer analysis. In addition, as suggested above, you may be stimulated to think about other ways in which computer processing can contribute to your own instructional objectives.

LIST OF TABLES

IMPROVING TECHNIQUES IN TEACHING ENGLISH FOR THE JOB

CHAPTER I: INTRODUCTION AND BACKGROUND

This handbook is addressed primarily to instructors and administrators in bilingual vocational training programs for limited English speakers. (Although the ideas presented here are oriented toward students from non-English language backgrounds, many can be modified and applied to native English speakers whose language skills need development). The handbook is designed to answer the following questions:

- How can English language skills needed for success on the job be taught more efficiently and effectively?
- How can the time spent on lesson planning be shortened, so that there is more time available to:
    - o improve the quality of teaching
    - o develop more relevant teaching materials
    - o provide for individual needs in instruction?
- How can the instructor or administrator with no prior experience use existing computer resources to accomplish the preceding goals?

Suppose you are a teacher independently teaching English as a Second Language (ESL) in a bilingual vocational training program, and your task is to teach the English skills needed in an occupational area that was unfamiliar to you (e.g., automotive repair, building trades, or dental assisting). Your major goal should be to coordinate the content of your English lessons with the content of the vocational instructor's classes. You might do this by tape-recording and transcribing his or her presentations. Next, you would need to make a card file of all of the vocabulary and grammatical structures used.

You could also make a card file of all of the words in any corresponding textbook, and compare these with the file of words used in class. Then you could sit down with the instructor and decide which words need to be taught, and what their priority should be.

While all of these procedures would be very helpful for the integration of English teaching and vocational training, they would leave little time for you to prepare lessons and teach classes. You will easily see how time-consuming and tedious the process is by trying to make a complete alphabetized card file of _one_ typical page of text or _ten minutes_ of a transcribed lecture, on an experimental basis. Fortunately, computers are designed to perform tasks just such as this, and can save hours of time that might otherwise be spent in copying and shuffling cards. In fact, they can do much more than this. Users do not even have to know anything about how a computer works to obtain such results -- no more than does anyone who receives a monthly computerized bank statement.

The computer is a _tool_, just as the typewriter, calculator, or photocopier is. You don't need to know how to use these tools yourself if there is someone either within or outside of your institution who is available for that purpose -- such as a typist, an accountant, or a photocopying service. A computer operator in your own office or a computer service outside can provide many of these types of analyses which will be discussed in this handbook. The purpose of the handbook is to explain what kinds of information the computer can provide, what you need to know to secure the information, and how this information can free you to spend more time improving the quality and effectiveness of your program.

The handbook, incidentally, has been field tested with instructors who had never used computer resources before. Their unanimous conclusion was that imagination was the only limitation on the potential uses for the information produced by the computer. We hope that you will agree.

The following section answers common questions about bilingual vocational training and the use of computer resources in teaching English for the job. The section incorporates some of the ideas presented in this introduction and, more broadly, provides a quick overview of the field for those previously unacquainted with it.

QUESTIONS AND ANSWERS

Q. What is bilingual vocational training (BVT)?

A. BVT is a program which utilizes the native language of the trainee for teaching subject matter and skills, while developing the trainee's ability to use and understand English needed for the job.

Q. What are the advantages of BVT?

A. BVT offers three major advantages over single-purpose vocational or English programs. First, it makes vocational training available to persons whose limited English skills have previously excluded them from traditional programs, thus extending equality of educational opportunity to a group that is very much in need. Second, this type of training takes advantage of trainees' existing language and cognitive skills; it does not ask trainees to wait for their English competence to reach an adequate level to begin training. Finally, the English teaching is geared to specific job needs. This tailoring process increases motivation to learn, and assures greater job success and employer satisfaction.

Q. Is bilingual vocational training cost effective?

A. Studies have shown that learning, job placement, and retention rates are much higher for BVT programs than in non-bilingual programs for the same types of trainees. These advantages compensate for any additional initial costs. Trainees typically repay the cost of training within three years after graduation through taxes on their increased earnings. The programs are also cost-effective in terms of reduction in the cost of public support.

Q. What other features are typical of BVT programs?

A. The following are typical characteristics of successful BVT programs according to a national study (Troike et al., 1981):

- All staff are bilingual
- Strong counseling support is provided
- Staff are highly committed
- Close community relations are developed
- English is related to vocational content
- Staff are sensitive to cross-cultural differences

Q. What is the role of job-related ESL in a BVT program?

A. Job-related ESL instruction is designed to support

functionally vocational instruction and job preparation. In contrast to other programs, English is not taught as an end in itself, nor is it isolated from job-directed program content. The English/ESL instructor works closely with the vocational instructor to coordinate job-related English lessons with vocational content and communication requirements of the job.

Q. How is English instruction integrated with the vocational content?

A. Many ESL instructors in a bilingual vocational training program monitor vocational classes to learn the subject matter and skills being taught to trainees. English and vocational instructors meet regularly to consult on vocabulary and other language needs to be covered in the English class. The English instructor visits job sites to observe and record language interactions typical of different jobs. Vocational class lectures and demonstrations may also be recorded and transcribed as a source of information for English lessons.

Q. How can computer resources help make English lessons more relevant and effective?

A. Text materials, lectures, and demonstrations can be typed (keyed) into a computer, and various types of analyses (alphabetical and frequency lists and indexes, concordances, word combinations, and parts of speech combinations and lists) can easily be produced. The instructor can use these computer analyses in preparing lessons, tests, and practice materials drawn from lectures, texts, and job interactions.

Through such computer use, the English instructor is better able to communicate with the vocational instructor and has more time to meet individual trainee needs.

Q. Couldn't these analyses be done by hand?

A. Yes, but it would require a full-time effort, would be repetitive, and would leave little time for anything else, including teaching.

Q. Is prior experience with computers required?

A. No; all you need to do is find a computer operation that can produce the analyses for you.

Q. How can an administrator or instructor learn more about how computer analyses can be used to improve English instruction?

A. Read this handbook.

CHAPTER II: SOME USEFUL TYPES OF TEXT ANALYSES

Given the proper programming, a computer can regroup and list the words of a text in a number of ways which are useful to teachers. This chapter presents and explains a selection of computer analyses to help you understand what types are available and how they function. The analyses were produced by one of the computer programs listed in Appendix C--the TXTPRO program developed at Georgetown University. Most of these lists in the first part of this discussion are based on a text in dental assisting. This section begins with the simplest lists and works up to the more complex ones.

A. Simple Lists

The most basic kind of list is simply a list of the words that occur in the text; these can be given in any of several formats. The usual order is alphabetic. The chief value of such a list is that it provides a checklist to which various notes or checkmarks can later be added by hand. You might want to have such a list to be able to mark those words that have already been taught and the number of times they have come up in the teaching materials. Or you can add other notes to show the relative importance of the word as a technical term; to mark words that can best be taught in connection with some other word; or list teaching aids to be brought to class. Such a list is essentially a scratch pad to help you keep track of information pertaining to each word.

Table 2.1 presents a sample of such a list. The sample begins with <u>effort</u> and ends with <u>entered</u>. It includes highly technical words, some of which you may not know, and many relatively common words which you might not think of putting into your lessons if the list had not pointed them out to you. Also presented are words that the student is likely to meet in the vocational training classes. You can discuss these words with the vocational teacher to find out what they mean and their importance to the specific vocational lesson being studied. Once identified and reviewed by the vocational teacher, they can become the basis of a sequence of ESL activities.

Table 2.1: Alphabetic Word List

EFFORT	EMBEDDED	ENDOCARDITIS
EGG	EMBRASURE	ENDOCRINE
EGGS	EMBRASURES	ENDODONTIC
EIGHT	EMBRYO	ENDODONTICALLY
EIGHTEEN	EMBRYOLOGY	ENDODONTICS
EIGHTH	EMBRYONIC	ENDOTOXINS
EITHER	EMBRYOS	ENDOTRACHEAL
EJECTING	EMERGE	ENDS
EJECTOR	EMERGENCE	ENERGETIC
ELABORATION	EMERGENCIES	ENERGY
ELAPSED	EMERGENCY	ENERGY-RELEASING
ELASTIC	EMERGING	ENERGY-YIELDING
ELASTICITY	EMERY	ENFLURANE
ELASTICS	EMINENCE	ENFORCED
ELASTOMERIC	EMISSION	ENGAGE
ELBOW	EMITTED	ENGAGES
ELBOWS	EMOTIONAL	ENGAGING
ELECT	EMPHASIS	ENGINEERING
ELECTRIC	EMPHASIZE	ENHANCE
ELECTRICAL	EMPIRIN	ENJOY
ELECTRICITY	EMPLOY	ENLARGE
ELECTROLYTE	EMPLOYED	ENLARGED
ELECTROLYTIC	EMPLOYEE	ENLARGEMENT
ELECTROMAGNETIC	EMPLOYER	ENLARGEMENTS
ELECTRON	EMPLOYMENT	ENOUGH
ELECTRONIC	EMPTIED	ENRICHED
ELECTRONS	EMPTY	ENSUES
ELECTROPLATING	EMULSIFICATION	ENSURE
ELECTS	EMULSIFIED	ENSURING
ELEMENT	EMULSION	ENTANGLED
ELEMENTS	ENABLE	ENTERED

Another type of list which is somewhat more useful is a reverse alphabetic list. Here the words are arranged alphabetically on the basis of how they end. In such a list all of the words having the same ending are brought together, and it is possible to pick out at a glance all of the words which have a particular ending, such as the *-ing* in the verb, *-'s* in a noun, or *-ness* (which makes nouns from adjectives).

The sample given in Table 2.2 includes some of the words that end in *-ed*. Most of them are the past tense (or past participle) of verbs. These suggest ways in which verbs may be taught, including differences in pronunciation (*incited*, *proved*, *fixed*) and differences in spelling (*curved*, *tested*, *pitted*, etc.). With such a list, you can group the verbs into lesson units depending on the nature of the lesson you wish to teach.

Table 2.2: *Reverse Alphabetic Word List*

INCITED	MANIFESTED	PROVED
ACCREDITED	SUGGESTED	IMPROVED
BENEFITED	DIGESTED	APPROVED
LIMITED	INGESTED	CARVED
UNITED	CONGESTED	OBSERVED
INHERITED	INTERESTED	RESERVED
DEPOSITED	ARRESTED	PRESERVED
HALTED	TESTED	CURVED
MELTED	REQUESTED	VIEWED
CONSULTED	INVESTED	REVIEWED
GRANTED	LISTED	ALLOWED
WARRANTED	ASSISTED	SWALLOWED
UNWANTED	TWISTED	FOLLOWED
INDENTED	EXISTED	HOLLOWED
CEMENTED	POSTED	BURROWED
COMPLEMENTED	ADJUSTED	RELAXED
SUPPLEMENTED	CRUSTED	WAXED
PRESENTED	FITTED	UNWAXED
REPRESENTED	SUBMITTED	FIXED
MISREPRESENTED	ADMITTED	AFFIXED
PREVENTED	EMITTED	MIXED
PAINTED	COMMITTED	DECAYED
POINTED	PERMITTED	DELAYED
APPOINTED	TRANSMITTED	SPRAYED
PRINTED	PITTED	CONVEYED
TINTED	ALLOTTED	SURVEYED
COUNTED	DISTRIBUTED	ALLOYED
MOUNTED	REDISTRIBUTED	EMPLOYED
UNMOUNTED	ATTRIBUTED	DESTROYED
NOTED	DILUTED	CRITICIZED
SINGLE-ROOTED	SUBSTITUTED	OXIDIZED
MULTI-ROOTED	CONTINUED	STANDARDIZED

B. Frequency Lists

A frequency list provides still more information. In such a list, all of the words of the text are listed along with a number which shows how many times that particular word occurs in the text. Frequency lists can be organized in one of three major ways. In the first, the words can be ordered on the basis of frequency, usually with the most frequent words at the beginning and the less frequent words towards the end, as in Table 2.3. This is called a frequency order list. A second way is to have the words arranged alphabetically, with each accompanied by its frequency number. Table 2.4 shows such a list. The third way is to arrange the words in reverse alphabetic order according to the ending of the words, again with the frequency number alongside each word. This is the reverse alphabetic frequency list as shown in Table 2.5.

1. Frequency Order List

The frequency order list is usually most interesting in its middle range. Words which are of extemely high frequency are usually of high frequency in all texts, and so must be taught no matter what the subject matter is. In any English text *the* is the word which occurs most often, closely followed by *a*, *of*, *in* and so on. Words of extremely low frequency may or may not be useful for teaching purposes, and the teacher must make individual decisions about each one. But if the text on which the frequency list is based is technical, then the bulk of the technical vocabulary will appear in the middle of the list, and can easily be picked out. Also in the middle of the list will be found a range of vocabulary which has been called "subtechnical." The subtechnical vocabulary includes those words that, while not themselves technical, are necessary if the technical vocabulary is to be used appropriately. For example, in a situation where <u>screwdriver</u> is a technical term, such vocabulary items as <u>turn</u>, <u>clockwise</u>, <u>counterclockwise</u>, <u>handle</u>, <u>shaft</u>, <u>fit</u>, <u>twist</u> are part of the subtechnical vocabulary needed to describe what a person does with a screwdriver.

Table 2.3: <u>Frequency Order List</u>

ITEM	FREQUENCY	ITEM	FREQUENCY	ITEM	FREQUENCY
THE	9792	OPERATOR	198	MOST	94
.	4571	PLACED	196	RESTORATION	94
OF	4049	SURFACE	193	RUBBER	94
,	3220	CAVITY	185	THEIR	94
AND	2605	HAS	182	BODY	93
TO	2254	WHEN	179	LEFT	93
A	1969	THESE	167	PRIMARY	93
IS	1861	TISSUE	165	PROCESS	93
IN	1676	CAN	161	THAN	91
$UI	1434	PREPARATION	159	MAXILLARY	89
$P	1245	HAVE	156	HANDPIECE	88
BE	1025	ALSO	154	OCCLUSAL	88
OR	890	ORAL	150	PULP	88
FOR	826	MOUTH	147	INSTRUMENT	84
ARE	812	ENAMEL	142	PLACE	84
(	737	MORE	142	ANESTHESIA	83
)	731	ONE	142	OVER	83
WITH	696	THEY	140	USUALLY	83
AS	574	TIME	138	DENTIN	82
ON	544	USE	136	POSITION	82
--	511	WATER	134	BEEN	81
BY	511	ALL	132	GINGIVAL	80
TOOTH	481	HIS	127	SURGICAL	79
MAY	458	SEE	125	TYPE	79
$H(	449	OTHER	122	PLACEMENT	78
THAT	449	MATERIAL	121	UNDER	78
$H)	437	HAND	120	GENERAL	77
IT	422	SURFACES	120	MANDIBULAR	77
TEETH	403	AREA	119	PROVIDE	77
DENTAL	393	FORM	119	OFFICE	76
AN	385	ITS	118	FOLLOWING	74
FROM	366	FIG.	117	INSTRUMENTS	74
PATIENT	356	:	115	KNOWN	73
$UA	339	DURING	109	BECAUSE	72
WILL	336	TREATMENT	107	MATRIX	72
WHICH	317	SUCH	106	PROCEDURE	72
AT	286	EACH	104	UP	72
THIS	283	CROWN	103	ANY	71
"	276	HOWEVER	102	FOIL	71
ASSISTANT	263	RIGHT	102	BONE	70
USED	258	DAM	101	1	70
NOT	257	MATERIALS	100	AMALGAM	69
;	255	GOLD	99	ARCH	69
$U(	245	THROUGH	99	DESIGNED	69
$U)	241	DENTIST	97	THERE	69
FILM	236	PATIENT'S	97	3.	69
SHOULD	235	FIRST	95	LINGUAL	68
MUST	221	SMALL	95	COTTON	67
INTO	214	TWO	95	LONG	67
IF	201	2	95	SOLUTION	67

Table 2.3 gives the beginning of a frequency order list. The words _tooth_ and _teeth_ are unusually high on this list because the text on which it is based is a text on dental assistance. You find many technical terms, however, as the list continues (_anesthesia_, _mandibular_, _amalgam_) as well as many subtechnical words (_body_, _instrument_, _procedure_). If it seems desirable, you may shorten a frequency list by having the computer skip words of very high frequency, words of very low frequency, or both.

2. Alphabetic Frequency List

The alphabetic frequency list brings together all of the words that have the same beginning and gives their frequencies. This is particularly useful for finding out how many times a given word occurs in all of its various forms.

Table 2.4: Alphabetic Frequency List

ITEM	FREQUENCY	ITEM	FREQUENCY	ITEM	FREQUENCY
EXPLICIT	1	FACES	3	FEARFUL	2
EXPLORATORY	1	FACIAL	57	FEARS	4
EXPLCRER	21	FACIALLY	5	FEASIBILITY	1
EXPLCRERS	3	FACILITATE	8	FEASIBLE	3
EXPLORING	2	FACILITATING	1	FEATURE	1
EXPOSE	10	FACING	5	FEATURES	2
EXPOSED	44	FACICAXIAL	2	FED	1
EXPOSING	9	FACICGINGIVAL	1	FECERAL	3
EXPOSURE	58	FACICLINGUAL	2	FEDERATION	2
EXPOSURES	8	FACICLINGUALLY	1	FEE	4
EXPRESSED	1	FACSIMILE	1	FFED	1
EXPRESSION	1	FACT	6	FEEL	4
EXREMELY	1	FACTCR	14	FEELING	6
EXTEND	12	FACTCRS	26	FEELINGS	2
EXTENDED	13	FACULTATIVE	1	FEELS	1
EXTENDING	7	FADES	1	FEES	2
EXTENDS	4	FAIL	1	FEET	3
EXTENSION	12	FAILURE	6	FELT	1
EXTENSIONS	3	FAIR	1	FENESTRATICNS	1
EXTENSIVE	10	FAIRLY	3	FERNLIKE	1
EXTENSIVELY	1	FALL	3	FERRIER	3
EXTENT	16	FALLING	2	FERRIER-TYPE	1
EXTERICR	2	FALLS	3	FERROUS	2
EXTERNAL	6	FALSE	1	FERTILIZATION	1
EXTIRPATED	1	FAMILIAL	1	FETAL	3
EXTIRPATION	2	FAMILY	8	FETUS	5
EXTRA	8	FAN-SHAPED	1	FEVER	5
EXTRA-ORAL	1	FAR	10	FEW	24

Table 2.4 is similar to Table 2.1, but includes frequency numbers. By adding the frequency numbers for *expose*, *exposed*, and *exposing*, you get a frequency number for all forms of this verb. If you include related words such as *exposure* and *exposures*, you get a figure for the whole family of words used in this text (129). Such a figure makes this group of words significant enough to call for inclusion in your lessons.

You will also be able to compare terms at a glance. For example, the uses of the singular, such as *factor* (14) and *extension* (12), and the plural, *factors* (26), and *extensions* (3), together total forty and fifteen respectively. This finding may change the emphasis to be given to each particular term in teaching. If, on checking other similar pairs, you find that the singular is usually much more frequent than the plural, this will give you an indication of the relative amount of time that should be spent on drilling on plural forms.

With some irregular forms, such as the verb to *be* (1025), *am* (2), *is* (1861), *are* (812), *was* (25), *were* (13), *been* (81), and *being* (44), some of the forms will not stand close to each other even in the alphabetic list. Since, however, we know what to look for, we can locate them more easily in the alphabetic list than in the frequency list. Clearly the present forms (*is*, *are*) are more needed in technical English than the past forms (*was*, *were*).

Some words may have the same form but be different grammatically (<u>use</u>, the noun, and <u>use</u>, the verb; of course the pronunciation differs, but the written form is the same), or they may be different in meaning as well (<u>lead</u>, the metal, and <u>lead</u>, to conduct). Differences of this sort do not become clear in lists such as those discussed here.

3. <u>Reverse Alphabetic Frequency List</u>

The reverse alphabetic frequency list brings together all of the words which have the same ending. By comparing the total frequency of words with particular endings, it is possible to decide on the amount of emphasis to be given to particular endings or to particular ways of forming words. To a limited extent, such a list can help locate verbs or nouns for the user. Many will appear with certain endings (-<u>ed</u>, -<u>ing</u>, -<u>ate</u>, -<u>ify</u>, etc.). Many forms with -<u>s</u> and almost all with -'<u>s</u>, -<u>tion</u>, -<u>ness</u>, etc. will be nouns. The ending -<u>ly</u> is usually the signal for an adverb, and the base to which this ending has been added is almost certainly an adjective.

The sample given in Table 2.5 shows some of the words which end in <u>-ing</u>. You can see that some are forms of verbs, some (<u>fiber-producing</u>, <u>decision-making</u>) are compound adjectives, and some (<u>misunderstanding</u>) are clearly nouns. This listing will remind you of the many ways in which this ending is used in English, and provides hints for addressing this issue in the classroom.

Table 2.5: Reverse Alphabetic Frequency List

ITEM	FREQUENCY	ITEM	FREQUENCY	ITEM	FREQUENCY
NUMBING	3	MISUNDERSTANDING	1	CONVERGING	1
REABSORBING	1	OUTSTANDING	1	GAUGING	4
TUBING	16	ENDING	3	REACHING	2
FACING	5	BENDING	4	FAR-REACHING	1
PLACING	52	ASCENDING	1	TEACHING	6
DISPLACING	1	DESCENDING	1	APPROACHING	2
SPACING	2	OFFENDING	1	ATTACHING	1
TRACING	1	RECOMMENDING	1	BLANCHING	1
PRACTICING	1	PENDING	1	BRANCHING	1
LANCING	1	DEPENDING	10	PUNCHING	6
ADVANCING	1	SENDING	1	SCRATCHING	1
INFLUENCING	1	EXTENDING	7	ETCHING	1
FORCING	3	BINDING	1	RETCHING	2
REDUCING	10	GRINDING	11	STRETCHING	3
PRODUCING	12	UNWINDING	1	ITCHING	2
DISEASE-PRODUCING	1	BONDING	1	TOUCHING	8
ENAMEL-PRODUCING	1	CORRESPONDING	4	WEIGHING	1
CEMENTUM-PRODUCING	1	SURROUNDING	9	COUGHING	1
FIBER-PRODUCING	1	CODING	2	SLOUGHING	1
CANCER-PRODUCING	1	REWARDING	1	INTERMESHING	3
BEADING	1	ACCORDING	22	ESTABLISHING	2
HEADING	1	RECORDING	2	POLISHING	27
LEADING	2	TOOTH-RECORDING	2	FINISHING	22
MISLEADING	1	OCCLUDING	1	FLUSHING	1
KNEADING	1	INCLUDING	13	BRUSHING	4
READING	8	EXCLUDING	1	TOOTHBRUSHING	10
SPREADING	2	BEING	44	CRUSHING	9
LOADING	3	WELL-BEING	4	THING	2
INVADING	2	SEEING	2	BREATHING	4
ADDING	1	BUFFING	3	TEETHING	1
SHEDDING	3	GING	1	SOMETHING	3
PUDDING	2	ENGAGING	2	CLOTHING	2
PRECEDING	3	HEMORRHAGING	1	SMOOTHING	1
SUCCEEDING	2	PACKAGING	1	SOOTHING	1
BLEEDING	8	DAMAGING	1	ANYTHING	3
AIDING	3	ENCOURAGING	5	EVERYTHING	2
AVOIDING	1	WEDGING	2	LEAKING	1
DIVIDING	1	DISLODGING	1	SPEAKING	1
PROVIDING	15	GAGGING	2	BREAKING	6
SHIELDING	2	FOGGING	2	FLAKING	1
ENERGY-YIELDING	1	BULGING	2	MAKING	21
WELDING	3	HANGING	1	DECISION-MAKING	1
BUILDING	10	CHANGING	4	TAKING	8
REBUILDING	2	OVERHANGING	3	PAINSTAKING	1
FOLDING	1	RANGING	3	BACKING	2
HOLDING	15	ARRANGING	1	SNACKING	1
MOLDING	1	BRINGING	1	PACKING	2
DEMANDING	1	SPRINGING	1	CHECKING	3
STANDING	2	SWINGING	1	CLICKING	1
UNDERSTANDING	6	EMERGING	2	NICKING	1

C. Word Combinations

While lists of single words can be useful, it is sometimes desirable to study combinations of words. Such combinations of words may be called collocations; for purposes of this handbook, they will be referred to as combinations.

The computer can provide lists of 2-word combinations, 3-word combinations, or, indeed, word combinations of any length. Usually 2- and 3-word combinations prove to be the most useful.

Not only can the word combinations be listed, but their frequency can be counted and the listings arranged in frequency order. They can also be arranged in alphabetic order by the first item in the word combination. Tables 2.6, 2.7, 2.8, and 2.9 show samples of how combinations can be listed.

Since the is the most frequent word in English, it follows that word combinations that include the word the are of high frequency. However, such terms will be less useful and interesting than other combinations which do not include such high frequency words. In order to shorten the list of word combinations, you may want to omit combinations that contain specified high frequency words.

Table 2.6 gives a sample of some of the most frequent 2-word combinations. The numbers on the left are simply for reference. The two words of the combination and the frequency of that combination in the text follow. Word combinations containing the have been omitted from the listing. You can readily pick out important technical word combinations (Nos. 12, 18, 31), important subtechnical vocabulary (9, 19, 23, 25) and high-frequency grammatical patterns (1, 3, 5, 9, 11, 29).

Table 2.6: 2-word Combinations in Frequency Order

OBS	WORD1	WORD2	FRQ
1	MAY	BE	249
2	IT	IS	166
3	SHOULD	BE	166
4	TO	BE	161
5	MUST	BE	133
6	SURFACE	OF	95
7	CAN	BE	91
8	WILL	BE	85
9	IS	PLACED	81
10	SUCH	AS	81
11	BE	USED	78
12	CAVITY	PREPARATION	75
13	KNOWN	AS	65
14	PATIENT	IS	64
15	USED	TO	63
16	AND	IS	61
17	AND	TO	61
18	RUBBER	DAM	60
19	PLACED	IN	57
20	THEY	ARE	55
21	TO	PROVIDE	52
22	USE	OF	52
23	SURFACES	OF	51
24	PER	CENT	50
25	PLACEMENT	OF	50
26	CHAIRSIDE	ASSISTANT	49
27	USED	FOR	49
28	USED	IN	49
29	IS	USED	48
30	IS	NOT	47
31	GOLD	FOIL	46

Table 2.7 gives a sample of 2-word combinations in alphabetic order by the first word. You can see immediately the importance of technical terms in which gingival is the first word, but other word combinations also stand out (2755, 2767, 2769).

Table 2.7: 2-word Combinations in Alphabetic Order

OBS	WORD1	WORD2	FRQ
2755	GAUZE	SQUARE	2
2756	GEAR-DRIVEN	HANDPIECE	2
2757	GEL	IN	2
2758	GEL	STRENGTH	2
2759	GENERAL	ANATOMICAL	2
2760	GENERAL	ANESTHESIA	28
2761	GENERAL	ANESTHETICS	4
2762	GENERAL	DENTIST	4
2763	GENERAL	FEELING	2
2764	GENERAL	PRACTITIONER	4
2765	GENERALLY	MEASURED	2
2766	GENERALLY	PREFERRED	2
2767	GENERIC	NAME	2
2768	GENERIC	NAMES	2
2769	GESTATION	PERIOD	2
2770	GINGIVA	AND	4
2771	GINGIVA	IS	3
2772	GINGIVA	TO	2
2773	GINGIVAE	AND	2
2774	GINGIVAL	AND	4
2775	GINGIVAL	AREA	4
2776	GINGIVAL	EMBRASURE	2
2777	GINGIVAL	ENLARGEMENT	2

Table 2.8 is a sample from the list of 3-word combinations in frequency order. If you look at Nos. 57, 71, 97, 98, you get insight into technical terminology. You get hints for teaching grammar from Nos. 51, 52, 54, 67, 75, 76, 89, 90, 91, 92. This last group underlines the importance of the passive voice in technical language.

Table 2.8: <u>3-word Combinations in Frequency Order</u>

OBS	WORD1	WORD2	WORD3	FRQ
51	IS	CAUSED	BY	8
52	IS	DESIGNED	TO	8
53	IS	ONE	OF	8
54	IS	USED	FOR	8
55	LEFT	HAND	AND	8
56	LINGUAL	SURFACE	OF	8
57	MAXILLARY	AND	MANDIBULAR	8
58	THEY	ARE	NOT	8
59	TO	AID	IN	8
60	TO	BE	RADIOGRAPHED	8
61	WILL	NEED	TO	8
62	ANGLE	FORMED	BY	7
63	AT	ALL	TIMES	7
64	AT	RIGHT	ANGLES	7
65	IN	WHICH	THERE	7
66	INSTRUMENT	DESIGNED	TO	7
67	IS	DERIVED	FROM	7
68	IT	CAN	BE	7
69	IT	IS	NOT	7
70	MAY	BE	UTILIZED	7
71	MESIAL	AND	DISTAL	7
72	OCCLUSAL	SURFACES	OF	7
73	OUTER	SURFACE	OF	7
74	PER	CENT	OF	7
75	SHOULD	BE	KEPT	7
76	SHOULD	BE	PLACED	7
77	THERE	IS	NO	7
78	TO	BE	PLACED	7
79	TOOTH	TO	BE	7
80	USED	IN	DENTISTRY	7
81	WHICH	THERE	IS	7
82	ZINC	PHOSPHATE	CEMENT	7
83	ALSO	BE	USED	6
84	AND	CAN	BE	6
85	AND	MUST	BE	6
86	AND	PLACES	IT	6
87	APPEARS	TO	BE	6
88	ARE	AS	FOLLOWS	6
89	ARE	DESIGNED	TO	6
90	ARE	KNOWN	AS	6
91	BE	STORED	IN	6
92	BE	USED	WITH	6
93	CHAIRSIDE	ASSISTANT	MAY	6
94	CHAIRSIDE	ASSISTANT	MUST	6
95	FINGER	AND	THUMB	6
96	FOR	USE	IN	6
97	GINGIVAL	MARGIN	TRIMMER	6
98	GOLD	FOIL	RESTORATION	6
99	HORIZONTAL	ANGULATION	IS	6
100	IN	COMBINATION	WITH	6

Table 2.9: <u>3-word Combinations in Alphabetic Order</u>

OBS	WORD1	WORD2	WORD3	FRQ
651	DENTAL	UNIT	AND	3
652	DENTAL	X-RAY	FILM	2
653	DENTAL	X-RAY	UNIT	2
654	DENTAL	X-RAY	UNITS	2
655	DENTIN	AND	CEMENTUM	2
656	DENTIN	FORMATION	IN	2
657	DENTIN	IN	RESPONSE	2
658	DENTIN	IS	FORMED	3
659	DENTINOENAMEL	JUNCTION	AND	3
660	DENTINOGENESIS	IMPERFECTA	IS	2
661	DENTIST	AND	HIS	3
662	DENTIST	IS	NOT	2
663	DENTIST	WILL	HAVE	2
664	DENTIST	WILL	INDICATE	2
665	DENTURE	BASE	RESINS	3
666	DEPEND	ENTIRELY	ON	2
667	DESIGN	IS	SPECIFIC	2
668	DESIGNED	FOR	CUTTING	2
669	DESIGNED	TO	BE	4
670	DESIGNED	TO	FIT	3
671	DESIGNED	TO	HOLD	2
672	DESIGNED	TO	PROVIDE	3
673	DESIRE	TO	EAT	2
674	DESTROY	OR	REMOVE	2
675	DESTRUCTION	OF	TEETH	2
676	DEVELOPING	PERMANENT	TEETH	2
677	DEVELOPMENT	BEFORE	BIRTH	2
678	DEVELOPMENT	IS	REFERRED	2
679	DEVELOPMENTAL	GROOVES	CROSS	2
680	DEVELOPMENTAL	PERIOD	IN	4
681	DEVELOPMENTAL	STAGE	OF	2

Table 2.9 is a sample from the list of 3-word combinations in alphabetic order by the first word. You can find important technical word groupings (2-word or 3-word) in Nos. 651, 652, 653-4, 655, 656, and 659. You can find common grammatical combinations in Nos. 666, 668, and 669-672. These can be worked into your lessons as you have the opportunity.

One chief advantage of lists of word combinations based on technical texts is that many of the technical terms are made up of groups of words, and so a list of word combinations quickly brings such technical terms to your attention. Perhaps you do not know them at all, or may know them but not recall them in preparing for class. With the list, already known expressions such as <u>dental floss</u> are immediately noticeable.

Unknown word groupings such as, perhaps, mucogingival junction can be discussed with the vocational teacher.

Another type of word combination that is easily found is the nontechnical combination. Nontechnical groupings are relatively prominent in written texts. Combinations such as according to, due to, in respect to, and by means of stand out clearly in text-based lists.

D. Concordances

A more complex type of listing, and one which can be extremely useful, is the concordance. This kind of listing is sometimes also called a KWIC, which stands for Key Word in Context.

In this listing all of the words of the text are arranged in an index column down the center of the page. Each word occurs in the concordance index as many times as it occurs in the text. To the left of each word is listed as much of the immediately preceding text as can be fitted between the word and the left margin. To the right of the index word is given as much of the text immediately following as will fit between the word and the right margin. The index word is usually set off by extra space. Since the index words are arranged in a column down the page, the extra space also forms a blank column down the page and so makes it easy for the eye to pick out the index word.

The amount of context that occurs in a line may vary. Most computers will print 132 letters or spaces on a line. Sometimes the line is shortened to 120 spaces or even to 80 spaces depending on the structure of the computer program or of the computer itself. The line is often shorter if a serial number is given. This shows the position of the index word in the text, and is usually at the right end of the line.

Concordances may list index words in any of several orders, but the most usual are alphabetic and reverse alphabetic order.

Table 2.10 gives a sample of a concordance in alphabetic order. It shows at a glance the many ways in which the word <u>out</u> is used in English, both as a separate word and as a prefix to verbs. You can also see the importance of <u>outer</u> in technical terminology; you would probably want to turn to the word <u>inner</u> in another part of the concordance to see if it is equally important.

Table 2.10: <u>Concordance of Text Items in Alphabetic Order</u>

) A PROPER RELATIONSHIP TO THE JAW AND TO EACH	OTHER	. THIS PHASE CONTINUES THROUGHOUT THE LIF
)WARD THE CUSP TIPS AND INCISAL EDGES . $P THE	OTHER	PORTION OF THE PULP IS MORE APICALLY LOCA
THE PERIODONTIUM $U) $H) $P TEETH , LIKE ALL	OTHER	ORGANS OF THE BODY , DO NOT EXIST INDIVID
IN TEETH THAT APPEAR TO BE QUITE NORMAL IN ALL	OTHER	RESPECTS . PULPECTOMY -- THE EXTIRPATION
T THEIR TARDINESS IS DEPRIVING THEMSELVES AND	OTHERS	OF TREATMENT . UP TO DATE AT ALL TIMES AN
EDUCATION COURSES WHEN THE DENTIST WILL BE	OUT	OF THE OFFICE SHOULD BE MARKED OFF . 4.
'S WHEN THE OFFICE IS CLOSED SHOULD BE CROSSED	OUT	. MINOR HOLIDAYS AND SCHOOL VACATIONS WHE
OF THIS POLICY AND TO SEE THAT IT IS CARRIED	OUT	. NEW PATIENTS SHOULD BE INFORMED OF THIS
) , OR STORAGE , FILE WILL BE MAINTAINED IN AN	OUT	OF HE WAY LOCATION . THIS IS USED TO STOR
'AKEN ? $P NEVER LEAVE RECORDS LYING AROUND OR	OUT	OF THEIR APPROPRIATE FILE SPACE . AS YOU
IE PERSON WRITING THE LETTER SHOULD BE WRITTEN	OUT	. IT CAN BE TYPED UNDER THE SPACE FOR THE
SPACE FOR THE SIGNATURE OR IT CAN BE WRITTEN	OUT	AS PART OF THE REFERENCE INITIALS . $H) $
)RD-KEEPING AND ACCOUNTS MANAGEMENT IS CARRIED	OUT	BY SOMEONE ELSE , IT IS HER RESPONSIBILIT
(D OF THE GROUP , SHE WILL ESTABLISH AND CARRY	OUT	PERSONNEL POLICIES . SHE WILL CONDUCT
MUST ALSO HANDLE ANY GRIEVANCES . $P TO CARRY	OUT	THESE RESPONSIBILITIES , THE OFFICE MANAG
(IMARY DENTITION . FADES , THE WRINKLES SMOOTH	OUT	AND THE BODY AND LIMBS ARE MORE ROUNDED .
SOLID STRUCTURES , THEN THEY BECOME HOLLOWED	OUT	, AND IN THIS FORM THEY SERVE AS MOLDS TO
WEAR CAN OCCUR ON INDIVIDUAL TEETH THAT ARE	OUT	OF NORMAL POSITION OR EXPOSED TO LOCAL TR
YERS . THESE ARE THE ECTODERM (ECTO- MEANING	OUTER	) , MESODERM (MESO- MEANING MIDDLE) AND
ARE TRANSVERSE , WAVELIKE RIDGES FOUND ON THE	OUTER	SURFACE OF THE ENAMEL ; TEETH , IS APPARE
(BER AND EXTEND IN A GENTLE $U(S-CURVE TO THE	OUTER	SURFACE OF THE DENTIN , WHERE IT IS COVER
(HITE LINE , LINING EACH SOCKET . $P THE DENSE	OUTER	CORTICAL PLATE PROVIDES STRENGTH AND PROT
-- ABNORMALLY FORMED DENTIN . ECTODERM -- THE	OUTER	PRIMARY EMBRYONIC CELL LAYER . EMBRYO --
VITY . DIFFERENTIATED FROM THE ECTODERM , THE	OUTER	LAYER OF THE PRIMARY EMBRYONIC CELLS .
) ARE MORE CONCERNED PHONE LINE , TRY TO SPACE	OUTGOING	CALLS SO THAT INCOMING CALLS MAY GET THRO
)INTMENT BOOK FOR SEVERAL WEEKS IN ADVANCE AND	OUTLINE	CERTAIN BASIC INFORMATION . THESE ENTRIES
, AND THIS INTERNAL WALL CLOSELY FOLLOWS THE	OUTLINE	OF THE EXTERNAL SURFACE OF THE DENTIN . $
MATRIX . IT IS THE FULFILLMENT OF THE PLANS	OUTLINED	AT THE STAGES OF HISTODIFFERENTIATION .

Table 2.11 shows a sample of a reverse alphabetic concordance, again focusing on the ending <u>-ing</u>. The use of this ending to form participles, adjectives, or nouns becomes obvious through this listing; you can thereby get a clear idea of the problems the student is likely to face when trying to make sense out of this ending in English.

Table 2.11: <u>Concordance of Text Items in Reverse Alphabetic Order</u>

TS . NEW PARTS MAY BE DIFFERENTIATED ,	CAUSING	SUPERNUMERARY CUSPS OR ROOTS . WITH A DEFIN
TION AFFECT THE INTERDENTAL PAPILLAE ,	CAUSING	FIRST INFLAMMATION , THEN BLUNTING ANF
INCOMING CALLS MAY GET THROUGH . WHEN	ORIGINATING	CALLS , BE SURE YOU HAVE ALL THE INFORMATI(
R FUSSURE MAY BE PRESENT AND PARTIALLY	SEPARATING	THE ADJOINING CUSPS . (A FISSURE IS A FAUI
BOTH AS AN ANCHORING LIGAMENT AND AS A	SEPARATING	TISSUE . HOLDS THE TOOTH IN PLACE AND
KEY TO A WELL ORGANIZED , EFFICIENTLY	OPERATING	DENTAL PRACTICE . A POORLY MANAGED BUSINES:
$U) $H) $P $U(TELEPHONE RECALL $U) .	CONTACTING	THE RECALL PATIENTS BY TELEPHONE IS AN
LINES ACROSS THE ENAMEL ROD STRUCTURE	REFLECTING	THE INCREMENTAL PATTERN OF GROWTH . INCREMI
ON . IT IS A STIPPLED , DENSE TISSUE ,	SELF-PROTECTING	IN FORM , AND IS FIRMLY BOUND AND RESILIEN
AL JUNCTION — THE PLANE FORMED BY THE	MEETING	OF THE DENTIN AND THE CEMENTUM . DENTINOEN/
IEL JUNCTION — THE PLANE FORMED BY THE	MEETING	OF THE DENTIN AND ENAMEL . DENTINOGENESIS
GOOD APPOINTMENT CONTROL PREVENTS LONG	WAITING	PERIODS . $P THE RECEPTION ROOM SHOULD BE
LENGTH OF HIS WAIT . IN SOME OFFICES ,	WAITING	PATIENTS ARE OFFERED A CHOICE OF COFFEE OR
OU SHOULD EXPLAIN THE SITUATION TO THE	WAITING	PATIENT AND ASK HIS COOPERATION . APPOINTMI
I LOCAL TRAUMA CAUSED BY HABITS SUCH AS	BITING	ON HARD OBJECTS , OR THE ENTIRE DENTITION I
ILARLY ARRANGED . THE TEETH OF CHILDREN	EXHIBITING	THIS CONDITION AND CHARACTERISTIC FOR EACH
OF THE LETTER , THE NAME OF THE PERSON	WRITING	THE LETTER SHOULD BE WRITTEN OUT . IT CAN I
R-PRODUCING CELLS . FISSURE -- A FAULT	RESULTING	AT THE JUNCTURE OF TWO D DEVELOPMENTAL CEN'
AE , CAUSING FIRST INFLAMMATION , THEN	BLUNTING	ANF CRATERING . $P ATTACHED GINGIVA EXTENDS
RSELF WITH ANY BUSINESS MANAGEMENT AND	ACCOUNTING	COURSES THAT ARE AVAILABLE TO HER . $H($U(
IAKE WISE USE OF THE RESCHEDULED TIME .	NOTING	THE PATIENT'S TELEPHONE NUMBER NEXT TO HIS
I ARE LOST AND THE FUNCTIONAL DEMAND OF	SUPPORTING	THE TEETH CEASES TO EXIST . ALVEOLAR SOCKE]
ZED 30 MINUTE APPOINTMENT -- NOR IS HE	WASTING	THE BALANCE OF A ONE HOUR APPOINTMENT . $HI
LIDES . $P A CROSS-REFERENCE FILE IS A	LISTING	, USUALLY ALPHABETICAL , TO ENABLE ONE TO
DESCRIBED AS THE ACT OF GRINDING OR "	GRITTING	" THE TEETH , ESPECIALLY DURING SLEEP . POI
. BRUXISM -- THE ACT OF GRINDING OR "	GRITTING	" THE TEETH , PARTICULARLY DURING SLEEP .
E . 3. MEETINGS -- REGULAR MEETINGS OR	CONTINUING	EDUCATION COURSES WHEN THE DENTIST WILL BE

The concordance has all of the advantages of the lists mentioned earlier, as well as others of considerable scope. With an alphabetic concordance it is possible to study any form of a word in the various ways in which it occurs in that text. You can see, for example, whether a verb is used in its transitive or intransitive form, or both. You will be able to see which forms of the verb are used more frequently. Further, where a form has more than one use or meaning, you can find the relative frequency of its various uses. With a reverse alphabetic concordance, grammatical categories can be studied.

For example, most verbs have a form which ends with -ed. This can be the simple past tense or it can be the past participle. When it is the past participle it may occur with have (have finished), in the passive voice with be (was finished); or as an adjective with a noun (the finished product). You may be surprised to find that in almost every technical text the passive use is easily the most frequent, followed by the adjectival usage. The use with have and the use as the simple past are surprisingly infrequent. Using the reverse alphabetic concordance, you will similarly see at a glance the various ways in which the ending -ing is used, and thereby gain a clear sense of what is important in the particular technical field.

Once you have decided what to teach, you can easily find examples in the concordance. The concordance will also provide realistic sentences for use in the classroom and pertinent ideas for classroom activities by using the context given in the concordance.

As stated before, it is important to consult with the vocational teacher to decide on key vocabulary to be taught. It is also important to work with the vocational teacher in deciding which sentences in the concordance are the most important to learn. The vocational teacher is best qualified to set the priorities for messages to be mastered and to adapt the textbook language of the concordance to the actual language of the classroom. The concordance provides you with much more language than can actually be taught, and is therefore best used as a rich resource for finding the messages most essential to the vocational lesson to be mastered. Once you, with the vocational teacher, have decided on the priorities for a lesson, demonstrated by a word list and corpus of sentences from the concordance using those words, you can independently prepare activities which result in mastery of the lesson.

It may seem that the concordance or KWIC index has all of the advantages of all of the other listings. This is essentially true. However, a word of caution is needed. An alphabetic and a reversed alphabetic concordance of an entire text may be large and heavy, especially if it is all bound together. It may therefore be advisable to check beforehand as to what weight of paper is involved. It may be preferable to request that the volume be separated and bound in a manageable format. Whatever the volume of the concordance, it will provide an invaluable reference for job-specific ESL lessons.

E. Word Segment Distributions

A word segment distribution is an index that shows which chapter or subdivision of the text a word occurs in, and how many times. Thus, if you know that Chapter 3 of the text is to be read at a particular time, such an index shows what words occur in it, whether they have already occurred in previous chapters, and whether they will occur again later in following chapters. Such a list can be a very useful tool for deciding on the vocabulary that you wish to emphasize at any given point in the course. It also helps in selecting the vocabulary that you will introduce for the first time, and the relative emphasis you will give to each item.

Table 2.12 shows the distribution of words in various sections of a text. The numbers on the left are for reference only. Next is the word and a number giving the total frequency of that word in the text. Finally, there are numbers giving the frequency of the word in each of the numbered segments of the text. You can see at a glance that certain words (199, adequate) are used frequently and are fairly evenly distributed through the segments. Adhesive (204), however, has almost all of its occurrences in segment 14; you could therefore teach adhesive later than adequate. If you know which segment the vocational teacher is preparing to take up next, you will know which vocabulary items the students will encounter in their vocational class.

Table 2.12: Word Segment Distribution

OBS	WORD	COUNT	SEG1	SEG2	SEG3	SEG4	SEG5	SEG6	SEG7	SEG8	SEG9	SEG10	SEG11	SEG12	SEG13	SEG14
193	ADDITIVES	3	0	0	0	0	0	0	0	0	0	0	0	0	0	3
194	ADDITONAL	1	0	0	0	0	0	0	0	0	0	0	0	0	0	0
195	ADDRESS	6	0	0	0	0	6	0	0	0	0	0	0	0	0	0
196	ADDRESSED	2	0	0	0	0	2	0	0	0	0	0	0	0	0	0
197	ADENOCARCINOMA	2	0	0	0	0	0	0	0	0	0	0	0	0	2	0
198	ADEQUACY	1	0	0	0	0	0	0	0	0	0	0	0	0	0	0
199	ADEQUATE	28	0	0	0	0	2	0	0	0	0	0	1	0	1	5
200	ADEQUATELY	1	0	0	0	0	0	0	0	0	0	0	0	0	0	0
201	ADHERE	4	0	0	0	0	0	0	0	0	0	0	0	0	0	3
202	ADHERING	4	0	0	0	0	0	0	0	0	0	0	0	0	0	2
203	ADHESION	5	0	0	0	0	0	0	0	0	0	0	0	0	0	5
204	ADHESIVE	14	0	0	0	0	0	0	0	0	0	0	0	0	0	9
205	ADJACENT	42	0	0	0	0	0	0	0	0	0	5	8	0	2	1
206	ADJOINING	6	0	0	0	0	0	0	0	0	0	1	1	0	2	0
207	ADJUNCT	1	0	0	0	0	0	0	0	0	0	0	0	0	0	0
208	ADJUNCTIVE	1	0	0	0	0	0	0	0	0	0	0	0	0	0	0
209	ADJUNCTS	2	0	0	0	0	0	0	0	0	0	0	0	0	0	0
210	ADJUST	3	0	0	0	0	0	0	0	0	0	0	0	0	0	0
211	ADJUSTABLE	1	0	0	0	0	0	0	0	0	0	0	0	0	0	0
212	ADJUSTED	3	0	0	0	0	0	0	0	0	0	0	0	0	0	0
213	ADJUSTING	3	0	0	0	0	0	0	0	0	0	0	0	0	0	0
214	ADJUSTMENT	6	0	0	0	0	0	0	0	0	0	0	0	0	0	0
215	ADJUSTMENTS	3	0	0	0	0	1	0	0	0	0	0	0	0	0	0
216	ADJUSTS	3	0	0	0	0	0	0	0	0	0	0	0	0	0	0
217	ADMINISTERED	27	0	0	0	0	0	0	0	0	0	0	0	0	1	0
218	ADMINISTERING	5	0	0	0	0	0	0	0	0	0	0	0	0	0	0
219	ADMINISTERS	1	0	0	0	0	0	0	0	0	0	0	0	0	0	0
220	ADMINISTRATION	26	0	0	0	0	0	0	0	0	0	0	0	0	3	0
221	ADMINISTRATIVE	16	0	0	0	0	14	0	0	0	0	0	0	0	0	0
222	ADMINSTRATIVE	1	0	0	0	0	0	0	0	0	0	0	0	0	0	0
223	ADMISSION	3	0	0	0	0	0	0	0	0	0	0	0	0	0	0
224	ADMITTANCE	1	0	0	0	0	0	0	0	0	0	0	0	0	0	0
225	ADMITTED	1	0	0	0	0	0	0	0	0	0	0	0	0	0	0
226	ADMITTING	2	0	0	0	0	0	0	0	0	0	0	0	0	0	0
227	ADOPTED	2	0	0	0	0	0	0	0	0	0	0	1	0	0	1
228	ADOPTION	1	0	0	0	0	0	0	0	0	0	0	0	0	0	0
229	ADRENAL	1	0	0	0	0	0	0	0	0	0	0	0	0	0	0
230	ADRENALIN	3	0	0	0	0	0	0	0	0	0	0	0	0	0	0
231	ADULT	9	0	0	0	0	1	0	0	0	0	1	1	0	1	0
232	ADULTHOOD	1	0	0	0	0	0	0	0	0	0	1	0	0	0	0
233	ADULTS	7	0	0	0	0	0	0	0	0	0	0	0	0	1	0
234	ADVANCE	8	0	0	0	0	5	0	0	0	0	1	0	0	0	2
235	ADVANCED	2	0	0	0	0	0	0	0	0	0	0	0	0	0	0
236	ADVANCING	1	0	0	0	0	0	0	0	0	0	0	0	0	0	0
237	ADVANTAGE	4	0	0	0	0	2	0	0	0	0	0	0	0	0	0
238	ADVANTAGEOUS	1	0	0	0	0	0	0	0	0	0	0	0	0	0	0
239	ADVANTAGES	8	0	0	0	0	0	0	0	0	0	0	0	0	0	5

F. Use of Grammatical Codes

For the types of reports that we have been talking about so far it is only necessary to have a text prepared for computer processing. The group of reports that we will now discuss also requires a computer dictionary.

A computer dictionary is simply a list of words with a specific kind or kinds of information given for each word. In an ordinary dictionary, the main information given is a description of the meaning of the word. But other kinds of information are also given: how the word is pronounced, where the word may be hyphenated, the part of speech of the word, the historical source of the word, the spelling of its inflected forms, and other similar information. The kind of information put into the computer dictionary depends on how the dictionary is to be used.

In connection with various kinds of research conducted at Georgetown University, a dictionary was prepared in which each word is assigned a grammatical code to show the way in which that word acts in the syntax of a sentence. These codes are simply groupings of letters that convey information. The first letter in each code shows the part of speech of the word (N for noun, V for verb, A for adjective, and so forth) and the subsequent letters in the code show various subclasses of that part of speech. A comprehensive explanation of the structure and meaning of the code is given in Appendix A.

It must be emphasized that the choice of these codes, while intended to be as generally useful as possible, has a particular purpose; any other code designed for other purposes can just as easily be used instead. These other codes can be simpler or more complex, depending on the purpose for which they are designed.

The codes supply the computer with information for use in processing texts. Suppose that all noun forms (N) that are countable (C) and plural (P) are given the code NCP. The computer program can bring together all countable plural nouns however the plural is formed. Thus, _men_, _women_, _feet_, _mice_, _shelves_, and _sheep_ can be brought together with _books_, _teachers_, _watches_, and _horses_, since all will have the same grammatical code.

The lists that we are about to discuss are similar in many ways to the lists already discussed, except for the fact that the ordering of the list focuses on the grammatical code and not on the spelling of the word.

These reports were also produced by TXTPRO, and are based on a text in automechanics.

1. Frequency Lists

Table 2.13 shows a sample that lists all of the grammatical codes and indicates the frequency with which they occur when assigned to the words of the text being analyzed. Words coded NCS (noun countable singular) appear more frequently (13.3 percent) than words with any other code. Also very frequent is the code TBCSP2. Here, however, the frequency is easily explained by the fact that it includes the determiner the. Indeed, this first type of code frequency listing needs further development if it is to be useful. Such development is supplied by the second and third lists as discussed below.

Table 2.13: Grammatical Code Frequency List in Alphabetic Order

CODE	FREQUENCY	CUM FREQ	PERCENT	CUM PERCENT
NBCS	178	765	4.433	19.054
NBCSA	1	766	0.025	19.078
NBCSP	14	780	0.349	19.427
NBS	89	869	2.217	21.644
NBSA	5	874	0.125	21.768
NBSY	1	875	0.025	21.793
NCP	172	1047	4.284	26.077
NCS	536	1583	13.350	39.427
NCSA	10	1593	0.249	39.676
NCSP	7	1600	0.174	39.851
NCSPY	5	1605	0.125	39.975
NE	10	1615	0.249	40.224
NEH3	2	1617	0.050	40.274
NR	38	1655	0.946	41.220
OAS	33	1688	0.822	42.042
OCI	4	1692	0.100	42.142
OCS	100	1792	2.491	44.633
TBCISP2	9	2119	0.224	52.777
TBCP1	14	2133	0.349	53.126
TBCSP2	354	2487	8.817	61.943
TBCS2	13	2500	0.324	62.267
TCP2	1	2501	0.025	62.291
TCS1Y	4	2505	0.100	62.391
TCS2	22	2527	0.548	62.939
TGP	1	2528	0.025	62.964
TGR	6	2534	0.149	63.113
TGS	2	2536	0.050	63.163
TG4	2	2538	0.050	63.213
VBD	1	2539	0.025	63.238
VBG	3	2542	0.075	63.313
VBN	1	2543	0.025	63.337
VBT	106	2649	2.640	65.978

The sample shown in Table 2.14 gives the grammatical codes in alphabetic order. Moreover, with them it lists, again in alphabetic order, all of the words of the text which have that code. The frequency of each word can be given so that the total frequency of all items having that particular code can be calculated.

Table 2.14: <u>Grammatical Code Frequency in Alphabetic Order by Code and Text Item</u>

CODE=DM1

TEXT ITEM	TOTAL OCCURRENCES
COMMONLY	4
COMPLETELY	2
CONSTANTLY	1
CURRENTLY	1
ELECTRICALLY	1
FIRMLY	1
FULLY	1
MECHANICALLY	1
PARTIALLY	2
PERIODICALLY	1
RAPIDLY	1
REGULARLY	1
RELATIVELY	1
SECURELY	1
THEREBY	1
THUS	1
USUALLY	1

TOTAL TEXT ITEMS WITH CODE DM1 = 22

CODE=DM17

TEXT ITEM	TOTAL OCCURRENCES
VERY	7

TOTAL TEXT ITEMS WITH CODE DM17 = 7

CODE=DM18

TEXT ITEM	TOTAL OCCURRENCES
GENERALLY	4
HOWEVER	1

TOTAL TEXT ITEMS WITH CODE DM18 = 5

Table 2.14 shows a listing that brings together various adverbs of manner (code DM). The numbers in the code divide the adverbs into groups depending on where they can occur in a sentence.

The third type of frequency list (Table 2.15) is similar to the second except that the list of words under each code is given in reverse alphabetic order.

Table 2.15: Grammatical Code Frequency in Alphabetic Order by Code, Then in Reverse Alphabetic Order by Text Item

CODE=VSN

TEXT ITEM	TOTAL OCCURRENCES
TAKEN	1
BROKEN	1
GIVEN	1
DRIVEN	2
DRAWN	5
THROWN	1

TOTAL TEXT ITEMS WITH CODE VSN = 11

CODE=VST

TEXT ITEM	TOTAL OCCURRENCES
BUILDS	1
DEPENDS	1
SURROUNDS	1
REDUCES	1
INDUCES	1
EXPLODES	1
ENGAGES	1
REACHES	3
ACCOMPLISHES	1
PUSHES	1
LIES	1
REQUIRES	1
INCREASES	3
RAISES	1
CLOSES	2
COLLAPSES	1
PASSES	2
COMPRESSES	1
LUBRICATES	1
CIRCULATES	1
DOMINATES	1
SEPARATES	1
OPERATES	1
INCORPORATES	1
CONTINUES	2
DRIVES	1
REVOLVES	1
INVOLVES	1
UTILIZES	1
WORKS	1
FREE-WHEELS	1
TRAVELS	1
OPENS	5
CONTAINS	1
BEGINS	1

Table 2.15 shows a listing that brings together certain verb forms. Those forms that can only be past participles of verbs have the code VSN. Since these verbs are irregular, this listing gives you a quick rundown of irregular verbs. The second part of the listing brings together the third person singular present of verbs; this list gives you a quick overview of the various forms of the verb ending from the point of view of both spelling and pronunciation.

Listings of this sort can help an English/ESL teacher decide on the emphasis to be placed on various parts of speech, and provide examples from the text of each type. In particular, the reverse alphabetic dictionary tends to draw attention to irregular endings, and to differences in the spelling of the regular endings.

2. Combinations of Grammatical Codes

The lists described here are essentially similar to those discussed previously, except in this case, combinations of codes are given, not combinations of words. To anyone familiar with the codes, certain grammatical patterns stand out. Thus, a 2-code combination in which the first code begins with the letters VB and the second is either VSDN or VSN is a clear indication of the use of the passive, since the first code indicates a form of the verb *to be* and either of the second codes indicates a past participle.

is repaired	are broken	was hauled
VBT VSDN	VBU VSN	VBD VSDN

If the second code is VSG, then the present continuous form is indicated, since the code VSG indicates the presence of the suffix *-ing*.

is repairing	was breaking	been hauling
VBT VSG	VBD VSG	VBN VSG

Listings of code combinations are usually more useful when three, four or five codes are examined together.

Table 2.16: <u>Text Segment Distribution of a 3-Code Combination</u>

OBS	PATTERN	COUNT	SEG1	SEG2	SEG3	SEG4	SEG5	SEG6	SEG7	SEG8	SEG9	SEG10	SEG11	SEG12	SEG13
46	VBT VSDN XF	7	0	0	1	0	0	1	0	2	3	0	0	0	0
47	VBT VSDN ZJP	9	0	0	1	1	1	1	0	2	0	1	1	0	0
48	VMUT VBU VSDN	8	0	0	0	0	2	0	2	0	0	1	1	1	1
49	VSDN P43Y ***	8	0	0	0	0	3	1	0	2	0	0	0	1	0
50	VSDN P43Y TBCSP2	9	0	0	1	0	1	2	0	2	0	1	0	1	1
51	VSDN P82 TBCSP2	6	0	0	0	0	1	1	0	1	0	1	2	0	0
52	VSDN P82 VSU	8	0	0	0	0	1	2	0	0	2	1	1	1	0
53	VSG TBCSP2 NCS	9	0	0	0	0	1	1	0	0	2	1	3	0	0
54	VST TBCSP2 NCS	8	0	0	0	0	0	2	0	0	0	0	1	0	0
55	VSU TBCSP2 NCS	8	0	0	0	0	0	0	1	0	3	0	2	1	0
56	XA TBCSP2 NBCS	6	0	0	1	2	0	2	0	0	0	0	0	0	0
57	XA TBCSP2 NCS	15	0	0	0	2	2	1	0	0	5	1	3	0	0
58	XF $ NCS	8	0	0	0	0	2	0	2	2	2	0	0	0	0
59	XF $ TBCSP2	22	0	0	0	0	1	3	6	1	3	4	2	2	0
60	XF $ ZCD	6	0	0	0	1	0	2	0	0	0	0	0	0	1
61	XF *** $	10	0	0	1	1	1	0	1	1	1	1	1	1	1
62	XF TBCSP2 NBCS	8	0	0	0	4	1	0	0	0	0	1	0	1	1
63	XF TBCSP2 NCS	11	0	0	0	0	2	2	1	3	0	1	0	1	0
64	XF ZCD TBCSP2	6	0	0	3	0	0	1	0	0	2	0	0	0	0
65	ZCD TBCSP2 NCS	13	0	0	1	1	0	5	0	0	2	0	1	0	1
66	ZJP *** NCS	6	0	0	1	0	2	3	0	0	0	0	0	0	0
67	ZJP TBCSP2 AUD	6	0	0	3	0	0	0	2	0	0	0	0	1	0
68	ZJP TBCSP2 NCS	24	0	0	1	4	1	5	0	4	1	1	2	2	0
69	ZJP TBCSP2 ZNV	6	0	0	0	1	1	2	0	0	0	1	0	0	0

While lists of this sort can be very useful to teachers preparing lessons, they are also extremely useful in checking technical textbooks in order to determine when structures are first introduced, how many times they occur, and how often they are repeated in the subsequent development of the text. Table 2.16 gives a sample of a 3-code combination listing giving the overall frequency and the frequency in each of the segments of the text. Nos. 46 and 47 are passive structures, No. 48 is a passive structure with a modal verb (VMUT), and so on.

3. Concordances with Grammatical Codes

The basic structure of the concordance with codes is the same as that of the concordance already described. The difference is that each word of the text occurs not only in a line of context but also with the addition of another line of the appropriate grammatical codes beneath all of the words in the first line including the index word. The concordance is then alphabetized on the basis of the codes attached to the index word. Thus, for example, all of the index words which have the code NCP are listed together, and within that list, the index words themselves are alphabetized. All of the words with the code NCS come next, and they are in alphabetic order within their own group.

Table 2.17 is a concordance in which each 2-line grouping is ordered on the basis of the grammatical code. The code ACO indicates the comparative form of an adjective; there were six such forms in the text. The code APO indicates the positive form of an adjective that can take the endings for comparative and superlative, but that does not have them in this use. Naturally, there were many more such examples in the text.

A concordance of this sort is an excellent reference tool for many of the more advanced stages of materials development, particularly when used in conjunction with other coded listings.

Table 2.17: Concordance of Text Items and Grammatical Codes in Alphabetic Order by Code, Then by Text Item

```
IL  IS  COOLED BY   PIPING IT TO  A            COOLER  UNIT IN  ONE OF   THE    RADIATOR TANKS .  $P A
3CS VBT VSDN   P43Y VSG    NR P82 ***          ACO     NCS  ZJP KAD P62Y TBCSP2 NCS     NCP   XF $  **

RATURES ,  LEANER FUEL MIXTURES ,              FASTER  OPERATING CHOKES ,  RECALIBRATED CARBURETORS ,
       XA ACO    NBCS NCP      XA              ACO     VSG       NCP    XA VSDN         NCP         XA

 AS    HIGHER ENGINE TEMPERATURES ,            LEANER  FUEL MIXTURES ,  FASTER OPERATING CHOKES ,
' ZCOPY ZAD    NCS    NCP             XA       ACC     NBCS NCP      XA ACC    VSG       NCP    XA

NEEDLE VALVE ,  AND FUEL WILL NO               LONGER  FLOW TO  THE    CARBURETOR .  WHEN FUEL IS  CONS
NCS    NCS   XA OCS NBCS VMUT TBCSP2           ACC     ZNV  P82 TBCSP2 NCS        XF ZCD  NBCS VBT VSDN

IIPPED WITH HELICAL GEARS TO  ASSURE           LONGER  LIFE AND QUIETER OPERATION .  ONE GEAR (   LOW )
ON     P91Y AUC     NCP   P82 VSU              ACO     NBCS CCS ACO     NCS       XF KAD NCS  XPC APO X

GEARS TO  ASSURE LONGER LIFE AND               QUIETER  OPERATION .  ONE GEAR (   LOW )   MAY  BE  OF
NCP   P82 VSU    ACO    NBCS CCS               ACC      NCS       XF KAC NCS  XPC APC XPC VMUT VBU P62Y

FERS TO  HOW WELL THE    ENGINE IS             ABLE  TO  DRAW IN  FUEL MIXTURE .  IT IS  DIFFERENT FROM
T    P82 DMQ ZADN TBCSP2 NCS    VBT            APC   P82 ZNV  ZJP NBCS NCS     XF NR VBT ALC       P55

AR .  SUN GEAR LOCK-OUT TEETH ARE              CLEAR  OF   THE    PINION CARRIER .  DRIVE IS  THROUGH
S  XF NCS NCS  NCS      NCP   VBU:             APO    P62Y TBCSP2 NCS    NCS     XF ZNV   VBT ZJP

ING OIL  .  $P LUBRICATING OIL  MUST           COOL  ,  CLEAN ,  SEAL AND LUBRICATE .  THE    $UA
    NBCS XF $  VSG         NBCS ZNV            APC   XA ZAV   XA ZNV  CCS VSU       XF TBCSP2 $

EITHER AN   OIL  BATH CLEANER OR  A            DRY  TYPE THAT UTILIZES SPECIAL PAPER AS    THE
ZOT    TCS2 NBCS NCS  NCS     OAS ***          APO  NCS  ZCT  VST      AUC     NBS   ZCOPY TBCSP2

' ARE  OF   THE    SINGLE PLATE ,              DRY  DISC TYPE .  ONE TYPE USES A   COIL SPRING AS    A
 VBU: P62Y TBCSP2 AUC    NCS   XA              APC  NCS  NCS  XF KAD NCS  ZNVS *** NCS  NCS    ZCOPY **

IUSTION CHAMBER DESIGN .  RICARDO ,  IN        EARLY  EXPERIMENTS ,  SET FORTH SOME    BASIC PRINCIPLES
        NCS     ZNV    XF NEH3    XA ZJP       APC    NCP         XA ZNV DL2   TBCISP2 AUO   NCP

AND AT  THE    SAME TIME PROVIDE AN            EASY  AND SMOOTH EXIT FOR THE    BURNED GASES .  $H( $U(
OCS P29 TBCSP2 AUR  VSDN VSU     TCS2          APC   CCS ZAV    NBCS P51 TBCSP2 VSDN   NCP   XF $(  $(

SEPARATES THE    MOVING PARTS WITH A           FINE  FILM .  AS    IT PASSES THROUGH THE    BEARINGS ,
/ST       TBCSP2 VSG    NCS   P91Y ***         APC   NCS  XF ZCOPY NR VST    ZJP     TBCSP2 NCP      XA

ERING ROD .  DURING THE    CRUISING AND        FULL  POWER RANGE ,  NO     FUEL IS  FED  BY   THE    ID
      NCS XF P48    TBCSP2 VSG      CCS        APC   NBCS  ZNV   XA TBCSP2 NBCS VBT VSDN P43Y TBCSP2 ZA

ED BY   FILLING THE    HOUSING PARTIALLY       FULL  OF   HEAVY TRANSMISSION GEAR OIL  .  AS    THE
  P43Y VSG     TBCSP2 NCS     DM1              APO   P62Y APC   NCS          NCS  NBCS XF ZCOPY TBCSP2

,  ARE  CLASSIFIED IN  MANY WAYS .  A          GOOD  ENGINE IS  DEVELOPED BY   HARD WORK ,  STUDY ,
XA VBU: VSDN       ZJP KAI  NCP  XF ***        APC   NCS    VBT VSDN      P43Y ZAD  ZNV  XA ZNV   XA

THE    AUTOMATIC TRANSMISSION INVOLVES         GREAT  DEAL OF   MATERIAL .  THIS CHAPTER HAS ENDEAVORE
TBCSP2 AUC       NCS          VST              APO    ZNV  P62Y NBCS     XF TBCS2 NCS    VHT VSDN

ORMAL ACCESSORIES ON  THE     ENGINE .         GROSS  BHP FIGURES ARE  WITH SOME    ACCESSORIES (   FAN
JG    NCP         ZJP TBCSP2  NCS    XF        APC    NF  NCP     VBU: P91Y TBCISP2 NCP         XPC NCS
```

The types of listings shown in this chapter can provide you with a starting point in preparing ESL materials and lessons that are specific to the vocational area for which the text analysis has been performed. As will be seen in the next chapter, it is not necessary to have all of these analyses to produce useful results. In fact, the simplest analyses may be the most immediately beneficial. However, the greatest payoff is found when using several different types of analyses together. Chapter 3 will demonstrate some procedures for doing this.

CHAPTER III: PROCEDURES FOR USING TEXT ANALYSES

A. Introduction

Now that you have a broad understanding of computer text analyses (as discussed in Chapter 2), the question arises: how can you make use of these analyses? This chapter will discuss ways to use different kinds of text analyses in developing language teaching activities related to specific vocational content. Some sample lesson activities are included for illustration.

Two goals of teaching English for the job should be kept in mind while preparing language lessons:

1. The lessons should contribute to the students' mastery of communication skills that are closely related to their vocational training and job requirements. ESL lessons should be organized around entire messages used in the vocational skills context. Mastery of particular grammatical or pronunciation features is distinctly secondary in importance to dealing with safety messages, processes, and directions, without which the worker cannot function.

2. The content of the English for the job program should be ACCURATE, AUTHENTIC, AND APPROPRIATE to the vocational area being studied. Clearly most ESL teachers, no matter how well-trained and experienced, do not themselves know the specific terminology or usages of auto mechanics, carpentry or electricity. Only the vocational instructor or someone intimately familiar with the field can judge the job accuracy, authenticity, and appropriateness of materials or activities.

As soon as you have some text analyses for your program in hand (methods for securing them are discussed in Chapter IV), you can begin to study the analyses in order to extract the pertinent information you will need for developing lessons. In doing so, you should keep in mind the type of the original text source(s) involved in each analysis in order to determine the relative importance of certain words or expressions and the context(s) in which they might be taught. You will also need to know what percentage of the text was sampled in order to judge the completeness of the analysis. If the printouts have come to you from another program or instructor, you will need to consult with these sources to obtain the needed information.

Here are some further considerations in working with the printouts:

1. Do not be overwhelmed by the size, bulk, or weight of the printouts. Most computer printouts take up a great deal of space. Learn to use them as you would any good reference material. You will never need to read them in their entirety any more than you would a dictionary. Become familiar with the contents so that you will know where to look for specific information.

2. Extract high frequency words and grammatical forms. Make a list of high frequency technical terms; these are words which will be used often and which must be mastered thoroughly. They can be used in most ESL lessons for the course. Look for common suffixes and word combinations to determine grammatical constructions to be taught.

3. Notice what is absent from the analyses. It will sometimes be just as important to note what does not appear as what does appear. For example, if the superlative form of adjectives is found to occur extremely rarely, then it would seem wise not to spend a great deal of class time teaching this form.

4. Consult with the vocational instructor(s) and job counselor. Ask them to critique your findings. Some high frequency items may not be important in their opinion while some crucially important items may not appear in the printout or may have a low frequency. Your original lists should be revised on the basis of their judgment.

5. Recognize the limitations of the computer analyses. Computers are good at mechanical kinds of analyses which would take you thousands of hours to perform. But they cannot identify appropriate social expressions, for example, or classify items or structures by function. They can be an aid, but they cannot replace your knowledge and judgment.

There is no automatic way to create lesson materials from the computer printouts--this will always depend, as it should, on your intelligence and imagination. The computer analyses are a resource which can give you time to prepare lessons, since they put at your fingertips a wealth of detailed information which would have been impossible to obtain otherwise.

B. How to Identify Job-related English Using Text Analyses

1. Using Frequency Ordered Word Lists

As a first example, let us suppose you are developing ESL materials for students in a computer "data entry" training program. One of the first things that you might do to determine the most important vocabulary for the job is to study the word frequency list, given in descending order of frequency. The word frequency list used here was based on an analysis of two textbooks on data entry. If you look at the words that occur more than 100 times each you will find 25 words that can be considered as high frequency (see Table 3.1). Of these, only eight are technical terms. (Note: in the table, the numbers indicate how many times each word appeared in the texts. The, for example, appeared 1,797 times.)

Table 3.1: Twenty-five Most Frequently Used Words in the Computer Analysis of Two Data Entry Texts

the	1797	in	335	will	199	data	161	are	130
a	497	and	334	on	197	field	159	if	124
to	478	card	255	record	190	or	136	this	107
of	442	be	207	program	177	for	135	when	106
is	374	key	206	you	173	disk	133	machine	101

The technical terms are card, key, record, program, data, field, disk, and machine. The most frequent words are the articles, the and a. The conjunctions and and or are on the list, as are will and various forms of the verb to be. The prepositions, to, of, in and on are represented here, too. This small portion of the word frequency list has already provided some useful information. As you can see, however, the computer has done little more than confirm your already existing intuitions about the important words. If you are the ESL teacher in a data entry program, you undoubtedly already know (or should know) that words like card, key, and disk are important. What you need to do, then, is to look at words appearing at the next highest level of frequency. In this example, words used between 20 and 100 times in the texts analyzed were chosen (see Table 3.2). You could arbitrarily select your own levels of frequency, depending in part on the size of the analyzed text.

Table 3.2: Words Occurring between 20 and 100 Times in the Computer Analysis of a Data Entry Text By Order of Descending Frequency

90-100	80-89	70-79	60-69
by	that	from	at
mode	it	punch	first
into	position	punched	search
column	as	each	cards
		with	which
		an	display
		press	one

50-59	40-49	30-39	20-29	
verify	number	correct	before	programmed
index	activate	keyed	copy	see
not	numeric	cursor	entered	status
character	punching	then	has	bed
all	station	track	left	but
set	only	blank	manual	called
characters	error	next	read	entire
keyboard	may	diskette	shift	feed
switch	must	have	unit	last
address	storage	information	enter	off
can	any	fields	job	out
	used	level	manually	programs
	operation	line	two	so
	operator	return	after	there
		being	buffer	under
		records	digit	also
		auto	skip	automatic
		function	special	hopper
		keys	columns	insert
		pressing	end	should
		displayed	use	time
			do	totals
			same	
			stored	
			alphabetic	
			automatically	
			been	
			current	
			new	
			screen	
			update	
			zero	

You can now study the two word lists (i.e., Tables 3.1 and 3.2), and try to subdivide the vocabulary into technical, subtechnical, and nontechnical categories. The technical words here would be those that are needed specifically for data entry (search, for example, does not at first glance look like a technical word, but does have a specific technical use in this field). Subtechnical words would be those that are needed in order to use the technical terms appropriately. Table 3.3 presents the results.

When the words were subdivided into the three lists they were kept in their relative order of frequency. Thus, card is the most frequent technical term and hopper is the least frequent (appearing about 20 times in the text). Position is the most frequent subtechnical term and totals the least frequent (again, appearing about 20 times). It is important to notice that these lists represent tentative divisions. You should consult the vocational instructor to ask if the lists conform to his or her perception of priority terms and concepts. The vocational instructor's judgment can revise priorities and classifications on the most accurate of computer-based lists.

2. Using Alphabetic Word Lists

The lists as they are now organized make other kinds of information easier to discover. Notice that the word program is the fourth most frequent technical term. If you look down the list of technical terms you will see that this word appears in other forms as well (programmed and programs). While you can compile this information

Table 3.3: Subcategories of Base Vocabulary for a Data Entry Text

Technical	Subtechnical	Nontechnical	
card	position	the	do
key	press	a	same
record	set	to	been
program	switch	of	current
data	number	is	new
disk	activate	in	zero
machine	station	and	see
mode	operation	be	but
column	correct	will	called
punch	information	on	entire
punched	level	you	last
search	line	or	off
cards	function	for	out
display	pressing	are	so
verify	entered	if	there
index	manual	this	under
character	read	when	also
characters	unit	by	should
keyboard	enter	into	time
address	manually	that	
numeric	digit	it	
punching	skip	as	
error	automatically	from	
storage	status	each	
operator	automatic	with	
keyed	insert	an	
cursor	totals	at	
track		first	
blank		which	
diskette		one	
fields		not	
return		all	
records		can	
auto		only	
keys		may	
displayed		must	
copy		any	
shift		used	
job		then	
buffer		next	
columns		have	
stored		being	
alphabetic		before	
screen		has	
update		left	
programmed		two	
bed		after	
feed		special	
programs		end	
hopper		use	

from the present list by inspection, you could get the same information more easily from the alphabetized word frequency list. Here, if you look up the word program, you will immediately see which forms of the word appear and with what frequency:

program	177
programmed	22
programmer	1
programming	3
programs	21
	224

Thus, while the word program itself appears 177 times in the text, the total number of times that it appears as a basic stem is 224 times.

If you look up data (the fifth most frequent technical term) on the alphabetized list, you will see that it is the only form and it appears 161 times. But punch, which is the 11th most frequent technical term when used as a word by itself, is more frequent than data when considered in all its various forms:

punch	75
punched	75
punches	12
punching	47
	209

This illustrates an important point. The individual computer analyses (in this case, descending order frequency listing and alphabetized frequency listing) are valuable when used separately, but they are even more valuable when used together.

3. Using Lists of 2-Word and 3-Word Combinations

As you saw in Chapter 2, the computer can produce analyses showing sequences or combinations of two or more words, listed either in descending order of frequency or in alphabetic order. This type of analysis can show information both about vocabulary and grammatical structure.

If you were to examine a printout of the twenty most frequent 2-word combinations found in a dental assistant text, the results would be as shown in Table 3.4.

Table 3.4: 2-Word Combinations in Dental Assistant Text, by Order of Frequency*

Word 1	Word 2	Freq.	Word 1	Word 2	Freq.
may	be	249	be	used	78
it	is	166	cavity	preparation	75
should	be	166	known	as	65
to	be	161	patient	is	64
must	be	133	used	to	63
surface	of	95	and	is	61
can	be	91	and	to	61
will	be	85	rubber	dam	60
is	placed	81	placed	in	57
such	as	81	they	are	55

* numbers refer to how often those two words were found together in that order in the text

Many of the 2-word sequences in Table 3.4 give information about frequently used grammatical structures, such as combinations employing the modal auxiliaries (<u>can</u>, <u>may</u>, <u>will</u>, <u>should</u>, and <u>must</u>). The most frequent sequence is <u>may be</u>. Similar sequences on the list are <u>should be</u>, <u>must be</u>, <u>can be</u> and <u>will be</u>. For examples of how these words have been used in the text, you will want to consult the printout containing the concordance for the text. If you don't have a concordance for this text, you can get further information on these words by consulting the analysis showing 3-word sequences or combinations.

As an illustration of how you could use this type of analysis to find examples of modal auxiliaries, in the dental assistant text analyzed here, the printout shows 28 different contexts in which <u>may be</u> appears. The most frequent is <u>may be used</u> (36 occurrences). Other sequences with <u>may</u> and <u>should</u> are:

may be caused	should be advised
may be designated	should be avoided
may be indicated	should be checked
may be prescribed	should be reviewed

Plentiful examples are listed for <u>must be</u>, <u>can be</u>, and <u>will be</u> as well. This "detective work" not only tells you something about the language of the text used for dental assistant training, but it also provides a multitude of examples for creating meaningful exercises.

The 3-word combination report for the auto mechanics text also provides information about the use of significant phrasal prepositions and conjunctions. For example, it is interesting that the third most frequent combination is the phrase *by means of* which occurred 47 times in the text. Other three word phrases which appear frequently are *as well as*, *in addition to*, *in relation to*, and *as soon as*. If you want to design some exercises with these "useful phrases", you can turn to the concordance for the text to see exactly how these expressions have been used.

Another important use of the list of 2-word combinations is that it will often reveal technical terms that are made up of word compounds. In Table 3.4, for example, you will note that *cavity preparation* appeared 75 times and *rubber dam*, 60 times. The original alphabetized frequency list of individual words shows that *cavity* appeared 185 times, *preparation* 159, *rubber* 94, and *dam*, 101. Thus, while these words don't always appear together, they do so with a high frequency. Other technical compounds in the analysis of 2-word combinations are *chairside assistant*, *oral cavity*, *oral hygiene*, *cotton pliers*, *general anesthesia*, and *spoon excavator*. This last term was used 24 times in the text. In consulting the alphabetized word frequency list for this text, you will find that *excavator* was used 26 times and *spoon* 24 times. This shows that *spoon excavator* is the significant sequence that should go in your list of technical vocabulary, not simply the individual words *spoon* or *excavator*.

In a list of most frequent 3-word combinations in an auto mechanics text, it is surprising how many are technical terms. Among other combinations, you will find:

transmission output shaft
rear axle housing
transmission input shaft
center thrust bearing
suction throttling valve
air conditioning system
internal combustion engine
master cylinder piston

These are all terms that you would want to include on your list of technical terms for this field. Further examination would reveal specific sequences to be included in the subtechnical vocabulary as well.

4. Using Word Segment Distributions

In the previous examples we assumed that the goal was to identify the technical and subtechnical vocabulary from an entire text, which would be taught over an entire course of study. You may, however, just want to make a vocabulary list for a single topic covered in one of the units in the vocational text. The computer can give you this type of information in a word segment distribution. The following example is drawn from a computer analysis of a basic electricity textbook, and will illustrate again the value of using more than one type of analysis at a time.

The word frequency list shows that the following are the ten most frequently occurring technical terms:

circuit	125
current	108
resistance	84
voltage	79

power	62
lamp	56
energy	49
electrons	43
volts	43
wire	39

It is clear, then, that these words must be taught. The only question that remains is _when_ they should be taught. You can obtain some preliminary information on this point by consulting the word segment distribution as seen in Table 3.5. This shows the number of times each word occurs in each segment of the text (each segment here corresponds to a chapter).

Table 3.5: _Word-Segment Distribution for Basic Electricity Text_

	Total	Seg. 1	Seg. 2	Seg. 3	Seg. 4
circuit	125	2	29	18	76
current	108	3	30	27	48
electrons	43	24	11	1	7
energy	49	21	8	11	9
lamp	56	0	0	1	55
power	62	5	8	31	18
resistance	84	0	45	16	23
voltage	79	0	32	23	24
volts	43	0	21	16	6
wire	39	4	11	0	24

The word _circuit_, which appeared a total of 125 times in the text, appeared most often in Segment 4 (76 times), and only twice in segment 1. This kind of information can be extremely important, and it is available for every word listed on the word frequency lists. Notice that the most evenly distributed word is _voltage_. The least evenly distributed word is _lamp_. It does not appear at all until segment 3, and then only once. Every other instance of _lamp_ is in segment 4. Clearly for that chapter, knowledge of this word is crucial.

If you were teaching a program utilizing this text, you would still need to check with the vocational instructor to learn how the topics were actually being sequenced in the vocational class, so that instruction could be appropriately synchronized. The word distribution analysis can provide an excellent basis for discussing questions of sequencing with the vocational instructor.

5. Using Reverse Alphabetic Word Lists

Suppose you wanted to devise some exercises which would teach students the use of a certain suffix. To do so, you could use the word frequency list that lists words alphabetically from the end of the word (reverse alphabetic list). Such a list produced from some food service manuals shows two full pages devoted to words that end in -*er*. In developing your exercise you would be able to choose -*er* words easily; these words will be ones that are actually needed in the job for which the students are being trained. Assume that you are interested only in teaching the use of the suffix -*er* that converts a verb to a noun representing an agent or instrument. When you look at the reverse alphabetic word frequency list you will find many examples of this usage, but you will also find words ending in -*er* that do not follow this pattern (e.g., *neater*, *wider*, *older*, which are comparative adjective forms). You can simply ignore these, as well as words which just happen to end in -*er*, such as *together*, *customer* and *shower*, and end up with over thirty such words. By narrowing this list down, perhaps including only those that appeared at least five times in the text, you would get the following list:

Table 3.6: Words Ending in -er in a Food Service Text

slicer	16	strainer	8
blender	5	scraper	13
worker	9	chopper	8
broiler	16	disposer	7
steamer	10	toaster	12
skimmer	5	mixer	12
cleaner	9	tenderizer	7
opener	5		

Some exercises using these words are presented in the next section of this chapter.

6. Using the Analyses Together

It should be emphasized that while each type of text analysis is useful in itself, additional insights can be gained by allowing the different types of analyses to play off of each other. Thus, while the frequency ordered word lists are useful for obtaining broad impressions of which vocabulary items are most relevant to the vocational area analyzed, the alphabetic list shows the various forms in which a particular root is used, and the word segment distribution identifies the order in which words need to be taught as they are introduced in the vocational class. Similarly, the lists of 2-word and 3-word combinations identify both important technical terminology and frequently used grammatical structures not seen by viewing words in isolation. Because the English language uses inflections in the endings of words for so many purposes, the reverse alphabetic word lists are particularly helpful for identifying vocationally relevant examples for whatever suffixes you would like to teach. Finally, the concordance will supply you with numerous examples of appropriate vocational context and messages to teach along with the vocabulary and grammatical patterns you have found using the other analyses.

The next section will provide examples of some lesson exercises and activities that can be designed using the text analyses.

C. Preparing ESL Lesson Activities

Once you, the ESL teacher, have some sound notions about the kinds of information you can expect to get from the computer analysis of text, you can turn your attention to the preparation of specific ESL lesson activities which will be coordinated both in content and time with the vocational program.

We cannot emphasize too strongly the importance of this coordination. You will be teaching a vocational language with which you are not thoroughly familiar, at least at the beginning. You will need the cooperation of the vocational staff to help you understand this language and set priorities for its teaching.

Of one thing you can be sure. The more you strive to teach the _authentic_ language of the vocation, the faster and better your students will learn English. The fact that they recognize this language as something they have immediate need for in training is enough motivation. The fact that they will practice what you teach in vocational class as well as in ESL class ensures the lessons will be learned.

The following is a sequence of steps that will be useful to follow whenever you are planning ESL activities to be coordinated with a vocational unit.

Step 1. Find out the general vocational topic to be studied. The vocational teacher may report: "We will be studying the unit on ____." or "I'll be teaching the material in Chapter 6."

Step 2. Using the word frequency list distributed by text segments, focus on items listed for that chapter or topic. Make a note as to how many times the word is used in the chapter.

Step 3. Concentrate on technical terms at first; non-technical and sub-technical items will emerge in later analyses.

Take the vocabulary list and the appropriate concordance to a meeting with the vocational instructor who will:

a. Select the words on your list which are most important to the topic.

b. Add words which may have been left out in the sampling strategy or which are preferred by the vocational teacher over the textbook language.

c. Select sentences in the concordance that illustrate appropriate uses of the vocabulary relative to the topic.

Step 4. Prepare a vocabulary list for the lesson for your own use and perhaps for the students' use.

Step 5. Analyze the vocabulary in context for aspects of grammar and pronunciation that need to be studied. The reverse alphabetic and word combination lists will be particularly helpful for this. If an analysis using grammatical codes is available it should be consulted.

Step 6. Prepare lesson activities appropriate to the vocational area of instruction and incorporating elements of ESL instruction. The following are general suggestions for the design of classroom exercises that draw on vocational content and emphasize student interaction:

a. Have the students give a non-verbal response (e.g. choosing, giving, pointing to an object) in response to questions, requests, or instructions.

b. Check comprehension of terms and their use in the context of a sentence by translating them into and out of the students' language.

c. Explain the meaning and use of the passive voice. If the trainees are Asian, you may note that the English passive does not connote misfortune as it does in many Asian languages.

d. Have students practice converting passive sentences to active sentences (e.g., convert *The intake valve is opened by the camshaft* to *The camshaft opens the intake valve*).

e. Have the students practice responding to context-relevant questions (e.g., *What opens the intake valve?* *When does the exhaust valve open?*).

The following discussion provides an example of how an ESL teacher in a house repair program might proceed through Steps 1 through 5 as they have been outlined above. Step 6 will be addressed in the context of several different vocational areas.

1. <u>Determine the specific vocational topic</u>.

The topic chosen from the general area of house repair is "painting and redecorating." In an actual job-related ESL class, the choice of a topic should closely parallel the topic that is being studied in the vocational class.

2. <u>Using the word frequency list distributed by text segments focus on items listed for the chapter or topic of interest</u>.

Given the subject of "painting and redecorating", you would want to consult the word segment distribution analysis for this topical area. There were 250 words in the chapter on this topic that occurred more than 10 times.

3. <u>Concentrate on technical terms at first</u>.

Of the 250 frequently occurring words in the chapter, 105 can be identified as technical or subtechnical words. Table 3.7 lists these words along with the number of times each one occurred. However, many of these words are also heavily used throughout the text. In order to identify the key vocabulary for this topic you should concentrate on the words appearing heavily or exclusively in this segment of the text. Table 3.8 shows the result of this selection process. You should review these lists with the vocational teacher to assure their authenticity and to determine if other terms should be included.

Table 3.7: Technical or Subtechnical Words in the Segment on Painting and Decorating

Word	Count	Word	Count	Word	Count
adhesive	13	finishes	6	roll	6
alkyd	10	floor	29	roller	16
application	12	flooring	16	rollers	5
applied	20	floorings	1	room	48
apply	24	floors	21	siding	11
applying	14	grout	11	size	12
area	35	inch	19	smooth	11
areas	7	inches	21	space	10
boards	10	job	19	spaces	4
bristles	11	ladder	15	strip	28
brush	32	latex	25	strips	11
carpet	49	length	26	studs	11
carpeting	16	line	20	surface	72
carpets	7	long	15	surfaces	12
ceiling	42	mastic	13	tape	19
ceilings	13	material	18	tile	50
cement	10	measure	11	tiles	74
coat	24	metal	12	trim	20
coating	4	nail	11	trowel	10
coats	8	nails	10	use	51
color	20	paint	183	used	45
colors	13	painted	25	using	16
concrete	15	painting	35	vinyl	26
cover	14	paints	36	vinyl-asbestos	1
coverage	2	panels	21	vinyl-coated	1
covered	8	paper	25	vinyl-faced	2
covering	11	paste	11	wall	87
coverings	11	pasting	3	wallpaper	15
cut	19	pattern	22	walls	49
doors	16	pile	13	wide	10
dry	24	plaster	11	width	18
edge	13	plywood	10	widths	8
edges	19	remove	19	windows	20
enamel	12	removed	11	wood	47
finish	23	remover	6	woodwork	10

Table 3.8: Words Appearing Mainly in the Segment on Painting and Redecorating

	Total in Segment	Total in Text
carpet carpeting carpets	72	72
pattern	22	32
strip strips	39	87
tile tiles	124	169
trim	20	46
vinyl vinyl-asbestos vinyl-coated vinyl-faced	30	36
wall walls	136	265
mastic	13	14

4. Prepare a vocabulary list for your own use and perhaps for the students' use.

It will be useful, particularly for frequently occurring words, to group the various forms of a word together. With the frequency ordered list you can identify frequently occurring words, and then use the alphabetic frequency list to find the various forms of the words that are used. Table 3.9 shows a list of words occurring more than 20 times for the topic of painting and redecorating. It gives frequently occurring root words along with derivatives of the root and the aggregate number of times the word occurs in all its forms. This makes it easier to assess which words will be important in designing lessons for the topic.

Table 3.9: Words Appearing More Than 20 Times in the Segment on Painting and Redecorating

Word	Count
apply (application, applied, applying)	70
area (areas)	42
brush	32
carpet (carpeting, carpets)	72
ceiling (ceilings)	55
coat (coating, coats)	36
color (colors)	33
cover (covering, coverings, covered, coverage)	46
dry	24
edge (edges)	32
finish (finishes)	29
floor (flooring, floorings, floors)	67
inch (inches)	40
latex	25
length	26
line	20
nail (nails)	22
paint (painted, painting, paints)	279
panels	21
paper	25
pattern	22
remove (removed, remover)	36
roller (roll, rollers)	27
room	48
strip (strips)	39
surface (surfaces)	84
tile (tiles)	124
trim	20
use (used, using)	110
vinyl (vinyl-asbestos, vinyl-coated, vinyl-faced)	30
wall (walls)	136
wallpaper	15
wide (width, widths)	36
windows	20
wood	47

Table 3.10 gives the less frequently used terms from the segment on painting and redecorating.

Table 3.10: Words Appearing Less than 20 Times in the Segment on Painting and Redecorating

adhesive	13	measure	11
alkyd	10	metal	12
boards	10	paste	11
bristles	11	pasting	3
cement	10	pile	13
concrete	15	plaster	11
cut	19	plywood	10
doors	16	siding	11
enamel	12	size	12
grout	11	smooth	11
job	19	studs	11
ladder	15	tape	19
long	15	trowel	10
mastic	13	woodwork	10
material	18		

5. Analyze the vocabulary for aspects of grammar and pronunciation.

By studying the list in Table 3.9 you will notice that words like carpeting and flooring appear beside carpet and floor. These word pairs provide a good opportunity to talk about the meaning of the -ing ending in this context. You can consult the reverse alphabetic frequency analysis for the entire text to find more words conforming to this pattern. A list of such words includes:

fencing	roofing
carpeting	scaffolding
covering	siding
weatherstripping	paneling
sheathing	

You can also check the 2-word and 3-word combination analyses to find common grammatical usages and vocabulary items that might be better taught as two or three word units. Such an analysis from the house repair text reveals the following:

2-word combinations:

can be
should be
may be
be sure
portland cement
masonry cement
storm windows
wall covering
furring strips
latex paint
alkyd paint

3-word combinations:

can be used
be sure to
be sure that
make sure that
make certain that
make sure you

Finally, by scanning the lists for pronunciation problems, you may note that there are a number of consonant clusters at the beginning of words (gr, br, str, and st). The alphabetic frequency list of words for the entire text will identify numerous other examples of these clusters. Some words identified in this manner are:

st-

stair
staircase
staple
steel
steep
stepladder
steps
stick
stiff
stool
store
studs
sturdy

str-

straight
streaking
stretch
strip
stripes
stripping
stroke
strong
structure

6. Prepare lessons and activities.

From the preceding presentation you should have a fair idea of how to sort through the various text analyses and use them in combination to extract meaningful and relevant vocabulary and grammatical patterns. The final step (Step 6) is the preparation of ESL lesson activities that are appropriate to the area of instruction. Using the information gained from the text analyses you can begin to generate whatever exercises, drills, dialogues, or strategies work for you in the classroom. You will soon notice during the design process that whatever teaching strategy you use can draw upon information gained through the analyses. However, the kind of strategy that you use is not as important as the language that it serves to teach.

On the next few pages you will see some sample lesson activities using material drawn from computer analyses of text. These lessons have been created from analyses of several vocational areas and demonstrate a variety of strategies.

a. House Repair

The foremost objective in teaching job-related English is to present relevant vocabulary in a context that conveys the appropriate meaning of the words. This objective may be augmented by a grammatical device to involve the student in the words presented. In the following exercise, the focus vocabulary taken from a concordance for house repair is underlined; the sentences provide an appropriate vocational context.

Exercise: Convert from Passive to Active

	Passive	Active (Command form)
Example:	The molding strips are attached to studs or furring strips.	Attach the molding strips to studs or furring strips.

1. A small trowel is used to install tile.
2. The mastic is spread on the wall and allowed to set for 24 hours.
3. A vinyl wall covering is never hung over old paper.
4. A paper that has a horizontal pattern can be used to make a room appear wider.
5. The tool is held at an angle so that the grout is forced into the spaces between the tiles.

As with other vocational areas, the subject of house repair is deeply dependent on the use of active verbs. A list of text-based verbs can be found in the reverse alphabetic list of words ending in -ed. The usual vocabulary-based activities for the chosen verbs can be easily augmented with a dialogue and a chain drill, as follows:

Exercise: Dialogue and Chain Drill (Complete Sentence Responses)

Focus: Did you? ____ Yes, ____

Super - Landlord

Landlord: Did you finish everything?

Super: Yes, I finished everything.

Landlord: Did you clean the hall?

Super: Yes, I cleaned it.

Landlord: Did you sharpen the tools?

Super: Yes, I sharpened them.

Landlord: Did you oil the hinges?

Super: Yes, I oiled them.

Landlord: Did you install the lock?

Super: Yes, I installed it.

Chain Drill - Place on blackboard

1. clean	6. oil	11. open
2. sharpen	7. call	12. unplug
3. lock	8. install	13. unclog
4. finish	9. drill	14. change
5. push	10. pull	15. protect

Note that the dialogue given above can be used either for vocabulary context or for work with related grammatical structures (here, the object pronouns are practiced). The dialogue also reinforces the pronunciation of -ed as the regular -d sound.

b. Dental Assisting

Vocabulary-based lesson activities can be developed for any subject area. A segment of a dental assisting text dealing with histology is a good case in point. An attentive ESL instructor in this field who has access to a reverse alphabetic list will immediately notice that terms which can be related to each other in vocabulary instruction can become the natural subjects of a follow-up grammar lesson, as the following shows:

<u>Exercise</u>: Changing Nouns into Adjectives

-al

pathology	pathological
physiology	
chronology	
clinic	
cervix	

-like

collar	collar-like
business	
web	
wave	
leaf	

-ic

system	systemic
embryo	
cyste	
microscope	

-ful

care	careful
help	
cheer	

-ary

the maxilla	maxillary
the saliva	

In this list, the natural clustering of the words suggests strategies for teaching vocabulary. For example, care, help, and cheer lend themselves to a lesson on waiting room etiquette. Drills for meaning, use in context, and translations of the terms can be followed by grammar instruction on the use of <u>-ful</u> for changing nouns into adjectives. Separate clusters (maxilla and saliva; system, embryo, cyste, and microscope) can be taught at different points both to present the new terms, and to reinforce the noun-into-adjective concept. The general concept of prefixes and suffixes can be summarized in an autonomous "affix" lesson well into the course.

c. <u>Chef's Assistant</u>

A primary concern of the ESL instructor in a vocational program related to cooking is the identification of major kitchen utensils and appliances. A visit to the program's lab kitchen or an observation of the vocational class in progress will readily show one common feature shared by kitchen items: the -<u>er</u> ending. As discussed earlier, a reverse alphabetic list produced from a standard food service manual may show two pages or more of terms ending with this suffix.

You may want to focus on just one use of the ending, changing a verb to a noun. This noun means "the thing or person that does what the verb expresses." (Example: "A <u>steamer</u> <u>steams</u> vegetables.")

<u>Exercise</u>: Make sentences using the words below:

	<u>Sample Words</u>		<u>Sentence</u>
(a)	broiler/broil/food	(a)	A broiler is used to broil food.
(b)	opener/open/cans	(b)	
(c)	strainer/strain/gravy	(c)	
(d)	tenderizer/tenderize/meat	(d)	
(e)	toaster/toast/bread	(e)	
(f)	mixer/mix/food	(f)	

Exercise: Answer the questions.

(a) Question: Is food cooked or mixed in a broiler?

Answer: Food is cooked in a broiler.

(b) Is food cooked or mixed in a blender?

(c) Do you use tenderizer for oatmeal or steak?

(d) Is the chef a worker?

(e) What do you think a disposer is used for?

(f) What do you think a scraper is used for?

(With exercises such as these, you can check on comprehension by the use of false examples, e.g., "Can you mix a spoon in a mixer?")

d. _Auto Mechanics_

While the teaching of extraneous features of grammar will be the exception rather than the rule, given the functional emphasis of vocational programs, there will be times in which such lessons will be essential to understanding. Significant phrasal prepositions and conjunctions are examples of grammatical devices which will be used in the broadest sense during a given course. Here, the 3-word combination list will be most useful. The list shows that _by means of_ appears the most frequently in an auto mechanics text for a total of 47 times. Other frequent 3-word phrases include _as well as_, _in addition to_, _in relation to_, and _as soon as_. The concordance supplements this information by placing the phrases in context.

Let's consider the expression by means of. Since the concordance is alphabetized by key words, the instructor can look up by, means, or of. Of the three, means might be the best one: by can be followed by many other words, and of can be preceded by other words, while means is most likely to be preceded by by. Turning to the page in the concordance where means is a key word, you will find examples in the text. Here are some of them:

1. The oil pressure is determined by means of an electric or a bourdon tube gage.

2. The input and output shafts are secured by means of snap rings and and bearing retainers.

3. The various linkage parts are connected by means of ball sockets.

4. Now hook the lid to a simple crankshaft by means of a connecting rod.

5. The breaker arm contacts the cam by means of a fiber rubbing block.

6. The breaker plate is rotated by means of a vacuum advance diaphragm.

7. The coolant may be channeled into an engine block by means of distribution tubes.

The term by means of is used to show how a process is completed (an instrument can complete a process). The resulting prepositional phrase answers the question "How?" As you can see from the above examples, most of the sentences in the text show that a process is accomplished by means of an object. When the second item is a process rather than an object, the term by is used instead of by means of ("The metal is treated by heating it.")

Possible exercise activities might include:

1. Use of how questions (elicit a short answer)

   - o How is oil pressure determined?
   - o How is the coolant channeled in an engine block?
   - o How are the linkage parts connected?
   - o How are the input and output shafts secured?

2. Conversion to informal speech register:

   - o What does the breaker arm do?
   - o What connects the linkage parts?
   - o What rotates the breaker plate?

3. Use in informal contextualized settings:

   - o Give me the (connecting rod) so I can (hook the lid to a crankshaft).
   - o I need the (   ) so I can (secure the crankshaft).
   - o You need the (ball sockets) so you can (   ).

D. Summary

This methodology of using text analyses for preparing job-specific ESL lessons was field-tested in several bilingual vocational training programs (BVTPs). Some of the exercises appearing in this chapter were created by ESL instructors in these programs, and we want to acknowledge their contribution. Activities and suggestions from the field of dental assisting were provided by Annette Zarian, an ESL instructor in the dental assisting BVTP at UCLA Extension. Ideas for the area of house repair were provided by Dennis Altfest, an ESL instructor in the building trades BVTP at Bronx Community College.

The examples in this chapter were presented to suggest the variety of applications that a resourceful ESL teacher can find for using the text analyses. They are certainly not an exhaustive treatment of the subject, nor are they intended to prescribe one set of teaching strategies over another. The essential point is that whatever methods you have found to be successful can be enhanced by the wealth of information available from the analyses. Our experience has been that even those most initially hesitant about using the analyses have quickly found uses that were not originally anticipated.

CHAPTER IV: OBTAINING TEXT ANALYSES

Assuming that you would like to perform some of the text analyses described in Chapters II and III, what steps do you need to take and what issues should you consider in order to accomplish them? This chapter attempts to answer these questions as well as to provide a basis for making cost determinations.

A. Steps in Creating Your Own Text Analyses

Figure 4.1 is a flow chart of the basic steps to be followed in obtaining and using computer analyses. The basic steps given in the flow chart are as follows:

1. Selection of materials
2. Text sampling
3. Data preparation
4. Data analysis

Each of these steps is discussed in detail below.

Step 1: Selection of Materials

Before you can begin to select materials, you must decide what kinds of language data you wish to have analyzed by computer and for what purposes. Language data selected may be either oral or written (see Appendix B for a detailed discussion).

Figure 4.1

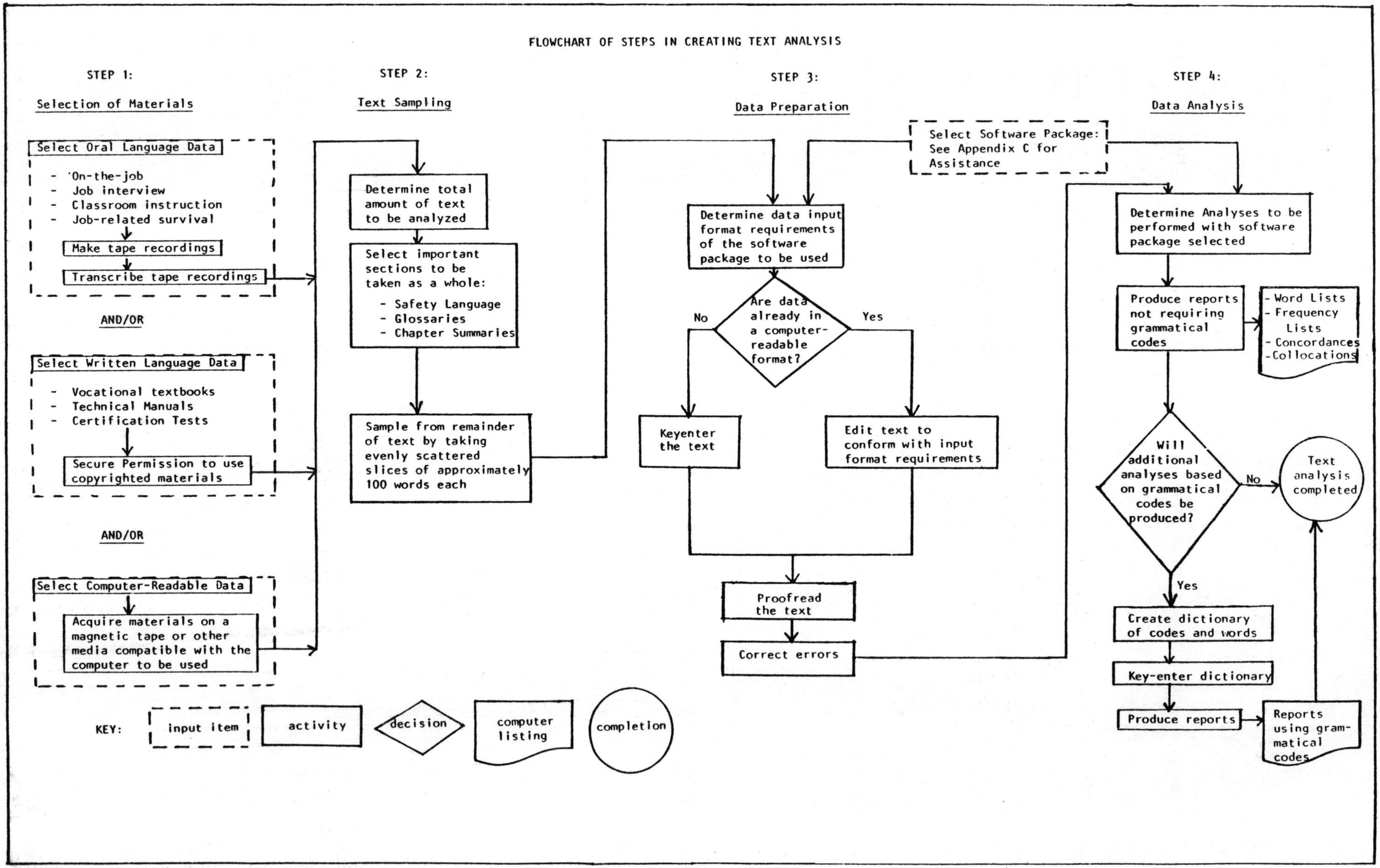
FLOWCHART OF STEPS IN CREATING TEXT ANALYSIS
STEP 1:
Selection of Materials
STEP 2:
Text Sampling
STEP 3:
Data Preparation
STEP 4:
Data Analysis
Select Oral Language Data
- On-the-job
- Job interview
- Classroom instruction
- Job-related survival
Make tape recordings
Transcribe tape recordings
AND/OR
Select Written Language Data
- Vocational textbooks
- Technical Manuals
- Certification Tests
Secure Permission to use copyrighted materials
AND/OR
Select Computer-Readable Data
Acquire materials on a magnetic tape or other media compatible with the computer to be used
Determine total amount of text to be analyzed
Select important sections to be taken as a whole:
- Safety Language
- Glossaries
- Chapter Summaries
Sample from remainder of text by taking evenly scattered slices of approximately 100 words each
Select Software Package: See Appendix C for Assistance
Determine data input format requirements of the software package to be used
Are data already in a computer-readable format?
No
Yes
Keyenter the text
Edit text to conform with input format requirements
Proofread the text
Correct errors
Determine Analyses to be performed with software package selected
Produce reports not requiring grammatical codes
- Word Lists
- Frequency Lists
- Concordances
- Collocations
Will additional analyses based on grammatical codes be produced?
No
Yes
Text analysis completed
Create dictionary of codes and words
Key-enter dictionary
Produce reports
Reports using grammatical codes
KEY:
input item
activity
decision
computer listing
completion

Oral language data include language used on the job, in job interviews, in classroom instruction, and as part of job-related survival. Written language data include language found in vocational textbooks, technical manuals, and certification tests.

Obviously the vocational area and the level of training will affect the priority of choices. In a short-term training program for an occupation that will make few literacy demands, the oral language used on the job, especially as it relates to safety, will be most important. In a BVT program that depends heavily upon trainees being able to read technical manuals and pass certification tests, the written language will assume greater significance. Further, the amount of reading that trainees will have to do in the program and on the job will also affect the choice of oral versus written language for computer analysis.

If you decide that you wish to include even some oral language in the material to be analyzed, you must tape-record actual speech in particular settings and transcribe it so that it can be input to the computer. It will be necessary for you to decide what settings and situations you wish to record (e.g., job, classroom, or elsewhere), and arrange to make recordings. Transcribing should be done by someone familiar with the situation, and the transcriptions should be carefully checked. While recording and transcribing oral language is time consuming and may entail additional costs, the results can more than justify the effort involved, since they will provide you with vocabulary and usages that may not be available in written form. Further, this material may be critical to the job success of the trainees. In addition, if both oral and written materials are used, it will be instructive to compare them.

In selecting written materials, you will want to choose materials that are most relevant to the needs of the trainees. These may include a variety of texts used on the job or in the training program. Since there is an expense in the preparation of materials for computer analysis, as well as in the analysis itself, you should carefully give priority to the choice of materials to be processed (unless you have an unlimited budget). You should also be careful to secure the permission of the publisher before using any copyrighted materials.

The selection of materials is an important area that calls for close collaboration between the English/ESL instructor and the vocational instructor. The vocational instructor is in the best position to decide what texts or parts of texts should have high priority for inclusion, while the English/ESL instructor is more likely to be sensitive to the linguistic complexity of texts (e.g., sentence and word length, number of subordinate clauses, and frequency order of vocabulary). This last factor may be relevant if alternative texts are available which differ in their appropriateness to the English reading ability of the trainees. Vocational instructors and administrators, even if they speak the trainees' language, may not be conscious of the language problems posed by particular training materials.

In some cases you may be able to find text material that is already in a form that can be read by a computer (i.e., can be input to a computer without any additional prior preparation; for example, this handbook was composed on a microcomputer). Some publishers now use computerized typesetting equipment, so that any materials they publish are already on a computer tape. Because much of the cost of having materials analyzed by a computer lies in the preparation of the text, it may be worthwhile to investigate whether a publisher can provide material in computer-readable form. Your choice of input language data, whether oral, written, or computer readable, should be made with the idea that it will become part of a resource library on which others may draw.

Step 2: Text Sampling

Once you have decided on the kinds of material (oral and/or written) that you want to have analyzed, you will need to decide how much of the text you wish (and can afford) to include. If the total amount of material is fairly small, probably all of it will be used. If, on the other hand, there is quite a bit of material, the time and cost involved may make it advisable for you to choose only a certain portion or portions for analysis. The following discussion of procedures for sampling text is based on written text, but everything that is said can apply to transcribed spoken language as well.

Suppose that it seems feasible to deal with only 25 percent of a text. The decision then becomes a matter of choosing which material to treat. Unless all of a particular quarter is of paramount importance for content reasons, it is generally better not to choose a single block of text, since this may very well leave out significant features found throughout the book.

It has been found in a number of experiments that, in order to be representative, the text to be chosen should be made up of short sections spread as evenly as possible through the text. Thus, if the text is 100,000 words long, and you wish to have 25,000 words processed, the optimum approach is to sample every fourth block of 100 words throughout the text. (Note that it is desirable for sections to begin and end with complete sentences, so a section might be slightly more or less than 100 words in length.) This procedure is more likely to produce a representative sampling of the entire text for analysis than other means.

Smaller samples, e.g. constituting 10 percent of the book, would follow the same pattern. However, if the sample is less than 25 percent of the text, it is desirable to supplement it with chapter summaries, glossaries, and any safety language that is highlighted in the text, so that nothing essential is missed.

Step 3: Data Preparation

Once you have chosen the text, you must enter it in the computer. There are a variety of ways this may be done. depending on the computer system, but all basically involve "keying" the data on a keyboard much like that of a typewriter. The simplest procedure is to key in the text in the same form and arrangement that you would use to type it on a piece of paper with a typewriter. Sometimes, however, you will need to make editorial changes in the text in order to make it easier for the computer to process the data.

For example, unless the text is specially marked, the computer may not distinguish between capital and lower case letters, since most computer input and output is entirely in capital letters. Similarly, some computer programs will require a special marker to indicate the beginning or end of sentences and paragraphs, and to identify abbreviations, underlining, or use of headings. Figure 4.2 shows a short passage of text that has been prepared for computer processing.

When you arrange for the use of a particular computer or computer service to process your data, you will need to ask what editorial conventions should be followed in preparing the text for processing. If you have been able to acquire a text that is already in computer-readable form, you may have to edit it so that it conforms to the conventions of the computer program to be used.

Figure 4.2: Sample of Text Which Has Been Prepared for Computer Processing

```
THROUGH LINES TO THE PUMP , WHERE IT IS EXPELLED THROUGH A              TXT01110
PULSATION DAMPENER CHAMBER THROUGH A FILTER TO THE CARBURETOR           TXT01120
FUEL BOWL .                                                             TXT01130
$P WHEN THE FUEL LEVEL IN THE BOWL RAISES , A FLOAT WILL BEGIN          TXT01140
TO CLOSE THE FLOAT NEEDLE VALVE . AT A PREDETERMINED LEVEL , THE        TXT01150
FLOAT WILL COMPLETELY SHUT THE FLOAT NEEDLE VALVE , AND FUEL WILL       TXT01160
NO LONGER FLOW TO THE CARBURETOR . WHEN FUEL IS CONSUMED AND THE        TXT01170
LEVEL STARTS TO DROP , THE FUEL PUMP WILL AGAIN FILL THE BOWL .         TXT01180
$P THE FUEL PUMP IS OPERATED ELECTRICALLY OR MECHANICALLY AND           TXT01190
MAY HAVE A FILTER EITHER ON , BEFORE OR AFTER IT .                      TXT01200
$H( $U( MIXTURE $U) $H)                                                 TXT01210
$P FOR EFFICIENT OPERATION AN ENGINE SUCH AS USED IN A PASSENGER        TXT01220
CAR REQUIRES AN AIR-FUEL RATIO THAT IS CONSTANTLY CHANGING .            TXT01230
THIS VARIED MIXTURE IS PRODUCED AND DELIVERED BY THE CARBURETOR         TXT01240
OR A FUEL INJECTION SYSTEM .                                            TXT01250
$P THE CARBURETOR CONTAINS A NUMBER OF CIRCUITS DESIGNED TO PRODUCE     TXT01260
A PROPER AIR-FUEL MIXTURE FOR ALL ENGINE NEEDS . THE GENERAL            TXT01270
CARBURETOR FUNCTION IS AS FOLLOWS :                                     TXT01280
$P WHEN THE ENGINE FIRST STARTS , THE CHOKE VALVE IS CLOSED .           TXT01290
VACUUM IS STRONG IN THE AREA BENEATH THE CHOKE VALVE , RESULTING        TXT01300
IN A VERY RICH MIXTURE BY REDUCING THE AIR SUPPLY AND INCREASING THE    TXT01310
WITHDRAWAL OF FUEL FROM THE BOWL . AS SOON AS THE ENGINE STARTS         TXT01320
, THE CHOKE PARTIALLY OPENS . AS THE ENGINE IS IDLED TO WARM-UP ,       TXT01330
FUEL IS FED INTO THE PASSING AIRSTREAM BY THE IDLE SCREW PORT . THE     TXT01340
CHOKE CONTINUES TO OPEN . WHEN THE ENGINE IS HOT , IT WILL BE           TXT01350
FULLY OPENED .                                                          TXT01360
$P WHEN THE ACCELERATOR OPENS THE THROTTLE VALVE , THE IDLE SCREW       TXT01370
PORT FUEL DELIVERY IS ASSISTED BY ADDITIONAL FUEL COMING FROM           TXT01380
THE IDLE PORT OR PART THROTTLE HOLES JUST ABOVE THE IDLE SCREW          TXT01390
PORT .                                                                  TXT01400
$P AS THE THROTTLE CONTINUES TO OPEN , THE AIR SPEED THROUGH THE        TXT01410
AIR HORN AND VENTURI INCREASES UNTIL FUEL BEGINS TO FEED FROM THE       TXT01420
MAIN DISCHARGE NOZZLE . AT VERY HIGH SPEEDS WHERE THE THROTTLE IS       TXT01430
WIDE OPEN OR WHEN THE CAR IS PULLING VERY HARD , ADDITIONAL FUEL        TXT01440
IS ADMINISTERED BY A POWER VALVE OR BY A                                TXT01450
METERING ROD . DURING THE CRUISING AND FULL POWER RANGE , NO FUEL       TXT01460
IS FED BY THE IDLE PORT OR IDLE SCREW PORT .                            TXT01470
$P TO ASSIST IN SMOOTH ACCELERATION , AN ACCELERATOR PUMP IS            TXT01480
PROVIDED TO FEED ADDITIONAL GASOLINE DURING ACCELERATION ONLY .         TXT01490
$P THE AIR AND FUEL MIXTURE IS DELIVERED TO THE COMBUSTION              TXT01500
CHAMBERS BY MEANS OF AN INTAKE MANIFOLD . THE TEMPERATURE OF THIS       TXT01510
MIXTURE IS SOMEWHAT CONTROLLED BY HEAT CROSSOVER PASSAGES , HEATED      TXT01520
AIR INTAKE AND , IN SOME CASES , BY RUNNING THE COOLANT THROUGH         TXT01530
PASSAGES IN THE INTAKE MANIFOLD . EVEN DISTRIBUTION OF FUEL TO ALL      TXT01540
CYLINDERS IS VERY IMPORTANT .                                           TXT01550
$H( $U( EXHAUST $U) $H)                                                 TXT01560
$P THE BURNED GASES ARE EXPELLED THROUGH AN EXHAUST MANIFOLD THAT       TXT01570
BOLTS ONTO THE BLOCK , OR ON THE CYLINDER HEAD , DEPENDING              TXT01580
ON THE ENGINE DESIGN . THE MANIFOLD SHOULD HAVE GENTLE BENDS WITH       TXT01590
NO BACK PRESSURE CAUSING KINKS . FROM THE MANIFOLD THE GASES PASS       TXT01600
THROUGH THE EXHAUST PIPE TO THE MUFFLER AND OUT THROUGH THE TAIL        TXT01610
PIPE . THE EXHAUST SYSTEM SHOULD MUFFLE ENGINE NOISE AND AT THE         TXT01620
SAME TIME PROVIDE AN EASY AND SMOOTH EXIT FOR THE BURNED GASES .        TXT01630
$H( $U( CARBURETOR TYPES AND KINDS $U) $H)                              TXT01640
$P THERE ARE DOWNDRAFT , UPDRAFT AND SIDEDRAFT CARBURETORS .            TXT01650
```

Step 4: Data Analysis

The kinds of analyses you can obtain will depend primarily on the computer program(s) available to the computer system you use. As part of the background to the development of this handbook, a survey was made of available computer programs ("software packages") that could be used for processing text data. A summary of the kinds of analyses each can perform is given in Appendix C. Not all programs can be used on all computers, so some compromise may be required if you cannot find someone with a computer that can run the program you would like to use.

Your choice of analyses to have done may be determined in part by what computer services are available, and in part by cost. For example, producing lengthier analyses like concordances (KWIC printouts) is generally more expensive than producing simple alphabetized word lists. If you wish to use grammatical codings for your text (see Appendix A), this will mean an additional cost in person-hours, as the codes must be assigned manually, either by you, a member of your staff, or the agency that does the computer processing. Once the coding is done, other useful analyses (discussed in Chapter 2) become possible. However, even if coding is not feasible, much valuable information and insight can still be gained from the less laborious and expensive types of analysis.

B. Cost Considerations in Creating Text Analyses

Since cost will probably be the major limitation on how much and how many types of analysis you will be able to have done, it is important to be aware of some of the cost factors involved in computer processing of text material. By knowing something about these factors, you will be better able to make realistic projections for program planning and budgetary purposes. Briefly, three major considerations need to be taken into account:

- Who will use the analyses?
- Who will do the analyses, and with what equipment?
- How much text can be analyzed economically?

Each of these points will be discussed in turn below.

1. Who Will Use the Analyses?

The principal point of this consideration is whether you can find any other potential users who might share the cost of the work with your program. If there are several programs using the same materials, the expense to each can be reduced by dividing the costs. At the same time, such sharing will allow for more complete analyses and text coverage. It is certainly uneconomical for several programs in different parts of the country to be duplicating the same work. Where available, it would be highly desirable if a central facility at the state, regional, or national level could carry out analyses and inform programs about the availability of analyses for particular text material.

2. Who Will Do the Analyses and with What Equipment?

Personnel and equipment are crucial points to consider because of the potential limitations they place on other decisions. As noted previously, you can do some textual analysis by hand with index cards and produce useful results. However, the variety and quantity of analyses that you could do this way are severely restricted. The choice of a computer or computer program will also impose certain restrictions.

The choice of which computer or computer facility to use may be the result of several considerations. Your organization or institution may already have a computer, or have access to the use of a computer, perhaps at favorable (non-commercial) rates. Before you can decide whether to use it or to look elsewhere, you will need to find out whether there is a computer program available for the machine that can be used to analyze your data.

The size and type of computer to be used is therefore a significant initial question. The computer, together with its associated peripheral equipment such as a printer, is collectively known as "hardware." Generally, three types of computers are in common use:

- Main-frame computers - These are the large, often multi-million dollar machines found in large universities, corporations and computer service bureaus. They can handle very large quantities of data easily.

- Mini-computers - These are mid-sized units often costing several hundred thousand dollars, and frequently used by colleges, schools, and small businesses. They handle smaller quantities of data, but at much lower cost.

- Micro-computers - These are desk-top units costing in some cases well below $10,000, and rapidly growing in use in schools, businesses, and even homes. While presently very limited in the amount of data they can handle, their capacities are expanding to rival those of mini-computers. Operating cost is very low.

There is a tradeoff between capacity to handle quantities of data and cost of operation. Although a main-frame computer might cost several hundred dollars an hour to operate, its ability to process enormous amounts of data in thousandths of a second can greatly reduce the overall cost. In addition, charges may be lower at certain times, such as at night. A micro-computer, by contrast, costs only pennies per hour to operate, but it may be severely limited in how much text it can handle at any one time--to the point that it may effectively be unable to produce the kinds of analyses discussed here. However, the state of the art of micro-computers is advancing so rapidly that some already have the capacity of mini-computers. Meanwhile, most micro-computers are too limited in their ability to store and manipulate the large amounts of text data required for the purposes considered here. Nevertheless, some restricted analyses are possible if programs exist or can be designed to produce the desired results.

The second major consideration is whether appropriate computer programs (software) are available for a given computer or computer facility to carry out the needed analyses. (A program is basically a set of instructions written to tell the computer what to do.) For certain models or brands of computers no software of this type may even exist, while for others the programs may have already been developed (See Appendix C). If they are not available at a particular computer facility at your own institution or elsewhere, they can usually be acquired and installed. However, it will probably require an additional expense to hire a programmer to do this.

If you wish to process large quantities of data (e.g., a fairly large textbook), you almost surely will need to use a main-frame computer. If such a computer is not readily available to you, it is possible that one of the organizations that originally produced the software listed in Appendix C may be willing to process your data on their computer for a reasonable fee. In fact, some organizations will not allow use of their computer programs except on this basis. You can write to these organizations explaining the size of the text to be processed and the way in which the analyses will be used, and you can usually receive information on the availability of their services, costs, and other important factors.

It will, of course, be helpful if you can find someone who is conversant in computer operations to provide advice on the relative costs involved in pursuing differnt options. Since text processing is a fairly specialized field, however, you should not expect most computer programmers to be closely familiar with the application of computers discussed in this manual.

3. How Much Text Can Be Analyzed Economically?

Since your budget will undoubtedly limit the amount of text that can be analyzed, it is important to identify the principal factors that must be taken into account in determining how much text can or should be chosen for analysis. The three major factors are:

- Cost of data preparation
- Sampling requirements
- Types of analyses desired

Cost of data preparation. Data preparation is the most costly step in the entire process, and will occupy about fifty percent of total expenses. As an estimate, data entry, verification, and correction will cost from 20 cents to 40 cents per card image (one 80-column line of text). As noted earlier, pre-editing in the form of inserting special markers for capitalization, sentences, and paragraphs will take considerable time. If you are including oral texts, you will have to allocate additional time and funds for recording and transcribing these before data entry can take place.

Sampling requirements. In sampling a text for purposes of language instruction, you need to use a sufficient amount of material to ensure a representative picture of the vocabulary and grammatical features of the text. Too small a sample can miss important items and give an erroneous impression of the characteristics of the text. In general, a sample of 100,000 words (actual words, punctuation marks, and any special editing symbols are counted as separate "words") of running text will provide from 85 to 90 percent of the technical vocabulary in a text, and 200,000 words will cover practically all of the vocabulary except for isolated proper names. If you are interested in grammatical patterns, a sample of from 40,000 to 50,000 words will provide about 95 percent of the major grammatical (syntactic) structures. These estimates are intended only as guidelines, since the actual amount of text required for an adequate sample will vary depending on the subject and the particular text.

Types of analyses to be performed. The cost of analyzing the data is affected directly by the amount of data being analyzed. While word lists are not strongly affected, the length (and hence the cost) of concordances rises dramatically as the text becomes longer. In a concordance, there is a line of print for every word in the text; therefore, a sample of 100,000 words will result in 100,000 lines of printout (1,666 pages). This is roughly one-half box of standard weight computer paper. If a concordance is produced along with other analyses, the total cost of the analysis phase will be about 60 percent of the cost of data preparation.

In summary, the main things you should consider in the creation of text analyses by computer are labor, equipment, the amount of text to be analyzed, and the kinds of analysis to be performed. The costs associated with each of these components can vary considerably depending on the setting in which the work is to be done. Data entry and verification is the most expensive step and will range between 20 cents and 40 cents per card image. The cost of the data analysis step depends upon who is doing the analysis, the computer facility to be used, the types of analysis to be done, and the amount of text to be analyzed. In budgeting for a project such as this, you should consult with someone familiar with computer costs and the use of the computer for text processing. It is ultimately your circumstances and the resources available to you that will define the amount and complexity of analyses you can perform.

C. Obtaining Existing Text Analyses

Suppose that you don't have the resources to create your own text analyses of the oral or written language used in your vocational training program. In such cases, is it possible to acquire analyses that have already been done for a particular vocational area? Some such analyses are available in the form of computer tape or printouts at the time of this writing (Spring 1982). For information on how to acquire one or the other, contact:

Lester Klein
U.S. Department of Education
(F.O.B. 6 - Room 1015)
400 Maryland Avenue, S.W.
Washington, D.C. 20202
(202) 472-1450

The examples of computer listings appearing in this handbook were taken from analyses of text materials for the following six vocational areas:

Auto Mechanics	Keypunching
Basic Electronics	House Repair
Food Services	Dental Assisting

The following publishers were kind enough to grant their permission to use the indicated texts as a source of written vocational language to be analyzed.

Goodheart-Willcox: South Holland, Illinois
Auto Mechanics Fundamentals, Stockel, 1978.
Exploring Electricity and Electronics, Howard Gerrish, 1981.

Mid-America Vocational Curriculum Consortium, Incorporated: Stillwater, Oklahoma
Food Service Production and Service: Student Materials, 1976.

Prentice Hall: Englewood Cliffs, New Jersey
Operating Data Entry Systems, Peggy Hanson, 1977.
Keypunching, Peggy Hanson, 1977.

Reader's Digest Association: New York, New York
Reader's Digest Complete Do-It-Yourself Manual, 1973.

W. B. Saunders: Philadelphia, Pennsylvania
Modern Dental Assisting, Hazel Torres and Ann Erlich, 1980.

To get further information regarding methodology of using text analyses and background research, contact:

Dr. Richard H. Naber
Office of Bilingual Education and Minority Languages Affairs
400 Maryland Avenue, S.W.
Washington, D.C. 20202
(202) 447-9227

The following types of analyses, as exemplified in Chapter 2, are available for all six vocational areas:

Simple Word Lists
Frequency Lists
Word Combinations
Concordances
Word Segment Distributions

For only the vocational area of auto mechanics are the additional analyses using grammatical codes available.

If your requirements are in a vocational area other than the six listed here, then you would probably want to consider creating your own text analysis by computer as outlined at the beginning of this chapter. If this is the case then we would encourage you to coordinate your activities with other groups that might benefit from this type of product. We have found a considerable amount of openness to the ideas expressed and encouragement from many groups in the development of this handbook. We are therefore hopeful that it will provide a starting point and guide in the coordination of teaching job-related English and vocational content areas.

APPENDICES

APPENDIX A

EXPLANATIONS OF THE GRAMMATICAL CODES USED IN THIS HANDBOOK

The grammatical codes used to illustrate the various reports appearing elsewhere in this handbook are part of a system developed at Georgetown University for a number of research projects.

It is not necessary to use these particular codes when reports using codes are desired. Any system of coding may be substituted, and the coding may be simple or complex depending on what the code user wishes to achieve.

However, since there is likely to be some curiosity as to the coding system illustrated here, explanations of the more frequently occurring codes are given in this appendix. Since the numbers occurring in the codes can safely be ignored, only the alphabetic parts of each code are listed. <u>It should be noted that these codes are designed to reflect the usage of each coded word as it occurs only in this particular text, and not as it may occur in English as a whole</u>.

The first letter of each code is the determinant, and this indicates the part of speech of the word coded. The following demonstrates the particular meaning of these determinants as used in this coding system:

A - Adjectives
C - Conjunctions
D - Adverbs
E - Existential Marker
J - Adjuncts
K - Numeratives
N - Nouns and Pronouns
O - Coordinators
P - Prepositions
T - Determiners
V - Verbs
X - Punctuation
Y - Idioms and Combinations
Z - Ambiguous Forms
$ - Clitics

The remaining letters in the code must be interpreted in accordance with the determinant they follow, and some letters have differing meanings depending on the determinant that begins the code. For example, the letter C in the code ACO indicates that the adjective in question (A) is in the comparative form; C in the code NCP indicates that the noun in question (N) is countable. The following gives examples of the more frequently occurring codes with an explanation of the meaning of each letter composing the code.

CODE	EXPLANATION	EXAMPLES
ACO	An adjective (A) with the comparative ending (C), and used both with nouns and as a verbal complement (O).	taller bigger greener
APO	An adjective (A), which takes endings when compared, but which occurs here in its positive form (P) and which is used both with nouns and as a verbal complement (O).	tall big green
ASO	An adjective (A), with the superlative ending (S), and used both with nouns and as a verbal complement (O).	tallest biggest greenest
AUO	An adjective (A) which does not take endings when compared (U), and which is used both before nouns and as a complement with verbs (O).	famous formal
AUR	An adjective (A) which does not take endings (U), and which is used only before nouns (R).	main former same
CA	A conjunction (C) which introduces relative (A) clauses.	that
CDF	A conjunction (C) which introduces adverbial (D) clauses which are conditional (F).	if
CDG	A conjunction (C) which introduces adverbial (D) clauses of concession (G).	though
CDT	A conjunction (C) which introduces adverbial (D) clauses of time (T).	when

CODE	EXPLANATION	EXAMPLES
D	An adverb (D).	very only
DL	An adverb (D) of place (L).	forward
DMQ	An adverb (D) of manner (M), interrogative (Q).	how
DM	An adverb (D) of manner (M).	quickly silently
DT	An adverb (D) of time (T).	soon never yesterday
E	The existential marker (E).	there
J	An adjunct or particle (these join with verbs to make compound verbs such as wash up, burn up, back up, carry over, make over, talk over).	up over
KAD	A number (K) which is a cardinal number (A) which is definite (D).	one ten
KAF	A number (K) which is a cardinal number (A), written in figures rather than spelled out (F).	14 32 176
KAI	A number (K) which is a cardinal number (A) which is indefinite (I).	several few
KOI	A number (K) which is an ordinal number (O) which is indefinite (I).	other
NBCP	A noun (N) which may be uncountable (B), or countable (C) and plural (P).	mechanics
NBCS	A noun (N) which may be uncountable (B), or countable (C) and singular (S).	gas
NBCSP	A noun (N) which may be uncountable (B), or countable (C) and either singular (S), or plural (P).	horsepower
NBS	A noun (N), uncountable (B), singular (S)	work time
NCP	A noun (N), countable (C), plural (P).	pistons
NCS	A noun (N), countable (C), singular (S).	piston

CODE	EXPLANATION	EXAMPLES
NCSA	A noun (N), countable (C), singular (S), likely to occur only as a modifier of another noun (A).	three-cylinder
NE	A noun (N) of the type usually called "proper" (E).	Smith Monday Chicago
NR	A noun (N) of the subclass pronoun (R).	it he they
OAS	A co-ordinator (O) of the "alternative" class (A), the second of two correlatives, or used singly (S).	or
OCS	A co-ordinator (O) of the "comprehensive" class (C), the second of two correlatives, or used singly (S).	and
P	A preposition (P) (Each preposition has its own 2-digit number which identifies it; this follows the P).	of by
TBCSP	A determiner (T) used with uncountable nouns (B) or with countable (C) nouns, whether singular (S) or plural (P).	the some any no
TCS	A determiner (T) used with countable (C) singular (S) nouns.	a an this that each
TGP	A determiner (T) which is possessive (G) and plural (P).	manufacturers'
TGR	A determiner (T) which is possessive (G) and pronominal (R).	its their
TGS	A determiner (T) which is possessive (G) and singular (S).	mechanic's
VBD	A verb (V), specifically "to be" (B) in the past (D).	was
VBG	A verb (V), specifically "to be" (B) in the present participle form (G).	being

CODE	EXPLANATION	EXAMPLES
VBN	A verb (V), specifically "to be" (B) in the past participle form (N).	been
VBT	A verb (V), specifically "to be" (B) in the third person singular present (T).	is
VBU	A verb (V), specifically "to be" (B) in the present other than third person singular (U).	are am
VHT	A verb (V), specifically "to have" (H) in the third person singular, present (T).	has
VHU	A verb (V), specifically "to have" (H) in the present other than the third person singular (U).	have
VMUT	A verb (V), modal (M) in either the third person singular (T) or any other form of the present (U).	must will can
VSDN	A verb (V), not an auxiliary (S), in the past (D) or in the past participle form (N).	opened provided checked
VSG	A verb (V), not an auxiliary (S), in the present participle form (G).	opening checking
VSN	A verb (V), not an auxiliary (S), in the past participle form (N).	broken given
VST	A verb (V), not an auxiliary (S), in the third person singular present (T).	flows travels
VSU	A verb (V), not an auxiliary (S), in the present, but not in the third person singular (U).	perform feed
VSUDN	A verb (V), not an auxiliary (S); either in the present, but not in the third person singular (U), or in the past (D), or in the past participle form (N).	shut cut
XA	A mark of punctuation (X) which may mark an end, or mark the boundary of a parenthetical insertion (A).	,
XF	A mark of punctuation (X) which marks an end (F).	. ; ! ?
XFL	A mark of punctuation (X) which marks an end (F) and may introduce a list (L).	:

CODE	EXPLANATION	EXAMPLES
XPOC	A mark of punctuation (X) which marks the boundary of a parenthetical insertion (P) and which may open (O) or close (C) such an insertion.	"
Z	An ambiguous form (Z). This code is always followed by at least two letters (determinants) which indicate the nature of the ambiguity. See the following examples.	
ZAN	An ambiguous form (Z) which may be either an adjective (A) or a noun (N).	patient
ZNV	An ambiguous form (Z) which may be either a noun (N) or a verb (V).	drive pass
ZOT	An ambiguous form (Z) which may be either a co-ordinator (O) or a determiner (T).	both
$	A clitic ($). Clitics are special markers introduced into the text to indicate how the words are arranged on the page of the original text, whether a word begins with a capital letter or is composed of capital letters, whether it is part of a title or heading, and so forth. All of the clitics introduced into the text begin with $. The grammatical code for any clitic is $ and a single letter, if the clitic indicates a characteristic of only the word immediately following. (See the following examples.)	
$P	Indicates beginning of paragraph.	
$H	Indicates a heading or title.	
$I	Indicates that the initial letter of a word is capitalized.	
$A	Indicates that all letters are capitalized.	
$U	Indicates underlining of text.	
$-(text $-)	If the special characteristic applies to two or more words, the clitic before the beginning of the word series will end with an open parenthesis. The clitic after the end of the word series will end with a closing parenthesis.	$U(AIR INTAKE$U)

APPENDIX B

METHODOLOGY OF TEACHING ENGLISH FOR THE JOB IN THE BILINGUAL VOCATIONAL TRAINING PROGRAM

A. The Philosophy of Bilingual Vocational Training (BVT)

Bilingual vocational training emphasizes the use of the trainees' native language to help them more rapidly acquire the knowledge and skills required for the job. At the same time, all BVT programs are concerned with teaching the specific English skills needed for success on the job. Some vocational areas may require relatively little English (chef's assistant, for example), while others will call for a high level of proficiency (e.g., dental assistant). The amount and kind of attention devoted to English language skills in BVT programs -- and the particular skills emphasized -- will thus vary by vocational area.

1. "General-purpose" ESL Does Not Work in Vocational Training

When bilingual vocational training had its beginning (1974), some of the early programs included a "general-purpose" ESL (English as a Second Language) curriculum which was not related to the vocational content of the programs. Administrators and teachers soon found that trainees progressed poorly; many became frustrated and dropped out, and those who completed the program had a difficult time communicating on the job. Out of this negative experience was born the realization of a need for close integration of the vocational and English/ESL components of the program.

2. The Need for a Functional Approach

These program findings were supported by research in second language learning, which has shown that language that is taught for its immediate functional value is remembered much better than language taught in isolation for its own sake. Since BVT programs are often of a fairly short duration, it is crucial that the English skills to be taught should reflect job demands and vocational skills content as closely as possible, and be presented in a functionally oriented fashion. This emphasis has resulted in the definition of what is referred to as "job-related ESL."

B. Job-related ESL

Prominent characteristics of job-related ESL include the following:

1. Pronunciation is de-emphasized in recognition of the fact that adults seldom completely master pronunciation of a second language. Perception, more than production, is stressed.

2. Grammar is de-emphasized, especially in early stages, and is presented cognitively. This recognizes that adult learners usually simplify grammar at first. Curriculum structure is instead based upon functional needs and program content.

3. Vocabulary is emphasized and related closely to vocational content and learner needs.

4. Memorization is minimized or eliminated except for routine formulas and safety language. Drill is based on meaningful communicative exchange.

5. The native language of the learner is used as an instructional tool to emphasize comprehension and the importance of meaning. The teacher should be bilingual if possible.

6. Reading is introduced early as an important skill and to reinforce oral practice. Writing is introduced as appropriate to the vocational needs of the job.

7. The class and curriculum are coordinated with the content and activities of the vocational component. The ESL instructor learns vocational content.

C. The Role of the ESL Instructor

It is important to realize that, within the BVT program, the ESL instructor occupies a service role supporting the goal of preparing trainees to function successfully in specific job settings. The purpose of the ESL instructor is, thus, to teach English language skills (including social interaction skills) needed by the trainees to follow vocational training and to survive and succeed on the job. The choice of material to be taught must, therefore, be practical and geared closely to the objectives and content of the vocational training.

D. Integration of ESL and Vocational Instruction

The need to integrate ESL instruction closely with vocational skills instruction has thus been recognized in practice. However, it is not always easy to bring about this integration. The ESL teacher frequently knows little about the vocational areas and as a result is not readily in a position to determine what technical vocabulary, grammatical structures, or usages are essential for trainees to learn, nor how to relate these meaningfully to specific skills or concepts. It is not enough to merely teach the pronunciation of lathe, debit, or molar, or to practice them in simple, fill-in-the-blank sentences ("The ____ is on the table").

It is therefore highly recommended that the ESL teacher learn as much as possible about the vocational area being taught, ideally by sitting in the vocational classes with the trainees (assuming the ESL teacher is bilingual). However, this may not always be practicable for such reasons as the following:

- The ESL or vocational teacher (or both) may be part time (a practice strongly discouraged), and unable to mesh schedules.
- The program budget may be insufficient to provide the instructor with free time for this purpose.
- A number of different vocational fields may be represented in the ESL class, making it difficult to learn about all of them.

Even in such circumstances, joint lesson-planning meetings between the vocational and ESL instructors can occur. The vocational teacher can often supply the ESL teacher with lists of important words and expressions to be practiced in the ESL class. However, since the vocational teacher is not familiar with second language learning and teaching isuues, he or she might well overlook certain "sub-technical" vocabulary or grammatical constructions that would be known to many native speakers but that would pose problems for second language learners. A good ESL teacher should be alert to this tendency. Of course, the ideal ESL instructor would be a vocational teacher who had been trained in second language teaching and was bilingual, thus understanding both knowledge areas and avoiding the deficiencies of each.

E. Types of Job-related English

Given the very limited time available in bilingual vocational training programs for English teaching, the instructor must determine trainees' learning needs and assign priorities to the language content and skills that must be taught to enable trainees to succeed in their training and to function successfully on the job for which they are being prepared. The language areas to be considered in making a needs assessment and planning a curriculum are as follows:

1. Safety Language

Since safety is always a prime consideration in vocational training, concepts, terms, and expressions concerned with safety must be given high priority in teaching. Usage in written form ("The use of X with Y may be hazardous and should be avoided," "Be careful not to come in contact with hot surfaces") will often differ considerably from oral usage on the job ("Don't ever use X and Y together!" "Watch out!" "Don't touch it! It's hot!"). Therefore both oral and written job usage must be taught.

Because of its critical importance, safety language should be translated into the trainees' native language to be used until English is thoroughly learned to make certain that the trainees fully understand the meaning and avoid injuries.

2. Specific Occupational Terminology

Occupational terminology includes a variety of terms with an emphasis on names for tools, names for persons in occupational roles, and labels for processes and concepts. It is important to emphasize that terms by themselves have little meaning: one could memorize a dictionary of electronics without really understanding how terms are used or what they mean. When new terminology is learned, the "semantic field" of existing concepts must often be restructured by comparisons with known objects and processes. At one level, for example, screws and nails may be contrasted; if bolt is taught, however, the concept of screw must be modified. The field of "screws" may subsequently be further subdivided by contrasting wood screws with metal screws. Similarly, learning warm and cool will require modification of the understanding of hot and cold.

Thus, before new terminology is introduced, it is important to know what terms have been learned previously, in order to know what kind of "cognitive map" trainees may have of the specific area.

3. On-the-job Language

There are many kinds of language usage (and social practices) on the job that trainees must learn if they are to function effectively and keep the job in which they are placed. Much of this usage ranges from the purely informal to the semi-technical and includes such things as: understanding directions, answering questions, asking for information, requesting permission, obtaining assistance, interacting with peers and supervisors, making excuses, and filling out timesheets or reports.

The appropriate language forms for carrying out these functions are not likely to be found in a textbook, but may be gathered in part by on-site observation and recording (including tape recording) of the language used. Scenarios can then be constructed and trainees can be taught to role-play using the appropriate language forms. For some, this may also include a certain amount of assertiveness training in the use of U.S. sociolinguistic rules of behavior.

4. Job-seeking Language

Since some training in job-seeking skills is part of bilingual vocational training programs, the job-related English instructor can and should work closely with the staff member responsible for that component to find out what language skills are involved. These can then be taught and practiced as part of the English-teaching (ESL) component. Typical skills would include making telephone calls regarding job openings, participating in mock interviews, and filling out employment applications. Part of job-seeking language is selling one's own abilities to a potential employer, a task that some cultural groups find exceedingly difficult.

5. Instructional Language

Instructional language is actually divided into three categories: classroom usage, textbook usage, and testing usage.

Classroom usage includes specific directions, signals of approval or correction, and vocabulary items (e.g., chalk, handout) that distinguish classroom routine (and sometimes individual instructors). These terms are often taken for granted with native English speakers but need to be taught explicitly to limited English speakers.

The second component is textbook usage. If a textbook written in English is used for vocational instruction, its distinctive features of language use need to be identified. Unique textbook expressions such as summary and review, may occur. More significant, however, is the "textbook prose" which differs from anything found on the job and may constitute an obstacle to trainees' learning. Sentence length and complexity, grammatical constructions, patterns or organization of information, and expectations regarding prior cultural knowledge may all present problems.

Finally, there is testing usage. Since they are not universal, test formats (e.g., true-false questions, multiple choice questions, and word problems) may be unfamiliar to many trainees. Also, test directions ("fill in the blank", etc.) may not be understood.

As already noted, samples of vocational text materials used in teaching or of technical manuals used on the job can be readily selected and analyzed by computer, as can samples of spoken language used in the vocational classroom or laboratory and on the job. The simplest procedure is to tape-record typical events over a period of several days, transcribe the recording, and submit the transcription for computer analysis. Previous research has shown that while _terms_ used orally on the job and in written texts are generally the same, _structures_ used in text materials, in lectures in vocational classes, and orally on the job tend to vary a great deal. Technical manuals in vocational education, though supposedly written at a fairly simple reading level, may be more complex and difficult to understand than material that is in theory at a more advanced level. The language used in tests may also present difficulties. For example, one widely used test, rather than asking _What is a carburetor for_? instead posed the complex question _What is the function of a carburetor in the operation of an internal combustion engine_?

Spoken vocational English will show some differences in terminology from that found in written texts, and usages may differ in various parts of the country. Differences may be matters of formality (i.e., wiggle versus move back and forth, or take off versus remove). They may reflect metaphorical usage, as in Jesus pin for cotter pin. In addition, trainees may, depending on the vocation involved, encounter a great deal of profanity on the job, and should be prepared to recognize it as such and interpret it accordingly.

F. The Goal is Communication

Methodologically, it is well to remember that ESL teachers in bilingual vocational training programs should emphasize the development of comprehension skills rather than language production skills. While both are important, comprehension skills can be developed more rapidly than production skills, and once developed, can form the basis for the growth of productive ability.

Since the time available for English teaching and practice in the vocational training program is so limited, it is important to seek an optimal balance between comprehension and production practice. The best way to determine what this balance should be is to make observations and recordings on a typical job site in order to find out how much and what kinds of language employees must use, and what they must be able to understand. That is, is it more important to be able to say, "I need a 3/4 inch hexagonal head screw," or is it more essential to be able to understand and respond if someone else uses the term? Practice in saying it does not necessarily assure rapid comprehension on hearing it.

Nor, significantly, does repeating terms <u>in isolation</u> lead to proficiency in actual use. Working through a substitution drill ("I need a 3/4 inch screw," "I need a 1/2 inch screw," "I need a 1/4 inch screw," etc.) contributes little to the ability to produce the appropriate utterance when needed on the job. Such a drill does not engage cognitive areas of the brain. Therefore, the practice tends to become a skill in itself, rather than part of a true communicative act.

Beyond very initial levels of introductory practice, then, drill activities should, to the extent possible, be embedded in realistic contexts or scenarios. If <u>comprehension</u> needs to be practiced, to use the earlier example again, students could be given an assortment of screws, and told, "I need a 3/4-inch screw", or "I need a 3/8-inch screw", and be required to respond by giving the teacher the appropriate screw within a timed limit (perhaps with an accompanying, "Here's one," or "Is this what you wanted?"). Both production and comprehension can be practiced by pairing students, and giving an object needing a screw in it to one student and an assortment of screws to the other student. The first student would have to determine the size of screw needed, ask the other for it, and on receiving it screw it into the object.

Since context and realization are so important to true communication, it would be well to carry out ESL instruction in the vocational lab where equipment is readily available, or arrange to have such materials placed in the ESL classroom as needed. Thus, everything from "Tap it a little on the side to straighten it up," or "Tomorrow is a holiday," to "I need a new typewriter ribbon," or "The belt on my sewing machine motor broke" can be practiced realistically and meaningfully.

The vocational areas with their vocabulary can be chosen depending upon the level of English ability of the trainees and the objectives of the program. In whatever is done, the following principles should be kept in mind:

1. Comprehension precedes production.
2. The order of importance in language learning is:

   Vocabulary (meaning)
   Grammar
   Pronunciation
3. Communication is the primary function of language, and language is learned best when used for meaningful communication.

The problem for the ESL teacher is, therefore, identifying the specific job-related English vocabulary and constructs that will be appropriate to the vocational training of the student, and teaching these in meaningful contexts.

For further information on methodology and practices in bilingual vocational training please refer to R. Troike, L. Golub, and I. Lugo, Assessing Successful Strategies in Bilingual Vocational Training Programs (Rosslyn, Virginia: National Clearinghouse for Bilingual Education, 1981).

APPENDIX C

BIBLIOGRAPHY OF ADP RESOURCES FOR BVT/ESL APPLICATIONS

A secondary object of this handbook is to identify existing automatic data processing (ADP) resources that are applicable to bilingual vocational training. A number of these resources or software packages are briefly characterized in the following pages. Each resource has been assigned an accession code and is arranged accordingly. The table that precedes these descriptions of the individual software packages provides an overview of the kinds of analysis which each may be used for. In this table the accession codes in the first column provide a reference to the individual bibliography entries that follow.

The various software packages do not perform all of the same types of analysis. Some may perform only one; others perform a series, usually in such a way that the user can select which will be used in a particular analysis. The types of analyses or tasks for which ADP resources may be useful follow. These are used to classify software packages in the chart (Distribution of ADP Resources) which follows.

I. Word Frequency Lists
II. Concordances
III. Statistical Analyses
IV. Readability/Stylistics
V. Test Development
VI. Record Keeping

This handbook has given the greatest attention to the first three types of analyses. They are the areas most appropriate to curriculum development, which is the most immediate need in job-related ESL instruction. Readability/stylistic measures may be useful for selecting appropriate instructional materials. However, many readability formulas are designed for children who are native English speakers, not adults learning English. Test development and record-keeping software may come into play at the curriculum evaluation stage. These software packages are more likely to be general teaching aids, independent of subject matter.

The following table may be used to identify ADP resources appropriate to your purpose by their accession codes. The accession code can then be used to look up the entry for the software package in the bibliography which follows. The bibliography identifies the name of the package, its developer(s), the programming languages used, and sources for further information. The information provided in the bibliography is the most complete and recent available, although there are some unavoidable omissions as well as dated material.

Distribution of ADP Resources

Accession Code	Analysis Types					
	Type I	Type II	Type III	Type IV	Type V	Type VI
DA001		X				X
DA002					X	
DA003				X		
DA004	X	X	X	X		
DA005			X			
DA007	X			X		
DA008				X		
DA009		X				
DA010		X				
DA011		X				
DA012	X					
DA013	X					
DA014						
DA015	X	X				
DA016	X					
DA017						
DA018			X			
DA020				X		
DA021	X					
DA022	X					
DA023	X	X				
DA024	X	X				
DA025	X					
DA026	X			X		
DA027			X			
DA028				X		
DA030				X		
DA031				X		
DA032				X		
DA034	X	X	X			
DA035	X	X	X			
DA036	X					
DA037			X	X		
DA038				X		
DA039				X		
DA040					X	
DA041					X	X

Analysis Types:

Type I - Word frequency lists
Type II - Concordances
Type III - Statistical Analyses
Type IV - Readability/Stylistics
Type V - Test Development
Type VI - Record Keeping

Accession Code	Analysis Types					
	Type I	Type II	Type III	Type IV	Type V	Type VI
DA042					X	
DA043						
DA044	X					
DA045			X			
DA046					X	
DA047	X		X			
DA049				X		
DA050				X		
DA051		X				
DA052						
DA053					X	
DB001	X	X		X		
DB003	X	X	X			
DB004				X		
DB005					X	X
DB006				X	X	X

Analysis Types:

Type I - Word frequency lists
Type II - Concordances
Type III - Statistical Analyses
Type IV - Readability/Stylistics
Type V - Test Development
Type VI - Record Keeping

PRIMARY ACCESSION CODE: DA001

TITLE OR NAME:
FAMULUS: A Documentation System

PROGRAMMING LANGUAGE(S) USED:
Not available

PLACE OF DEVELOPMENT:
Pacific Southwest Forest and Range Experiment Station in Berkeley, California

PREVIOUS RELEASE DATE:
1969

FOR FURTHER INFORMATION, CONTACT:
Dahlin, Gerry M.
Control Data Corporation
4201 Lexington Ave.
North Arden Hills, MN 55112

PRIMARY ACCESSION CODE: DA002

TITLE OR NAME:
GENRATE: A Means of Sentence Synthesis

PROGRAMMING LANGUAGE(S) USED:
PL/1

PLACE OF DEVELOPMENT:
Georgetown University

PREVIOUS RELEASE DATE:
1978

PROPRIETARY RIGHTS HELD BY:
Bernstein, Pamela
Georgetown University
Washington, D.C. 20057

PRIMARY ACCESSION CODE: DA003

TITLE OR NAME:
SPANFRQ: Automatic Morphological Analysis of Spanish

PROGRAM CREATOR(S)/DEVELOPER(S):
McCallum-Bayliss, Heather

PROGRAMMING LANGUAGE(S) USED:
PL/1

PLACE OF DEVELOPMENT:
Georgetown University

PREVIOUS RELEASE DATE:
September, 1980

PROPRIETARY RIGHTS HELD BY:
Linguistics Department
McCallum-Bayliss, Heather
Georgetown University
Washington, D.C. 20057

PRIMARY ACCESSION CODE: DA004

TITLE OR NAME:
TXTPRO: A Text-Processing Package

PROGRAM CREATOR(S)/DEVELOPER(S)
Shaefer, Leonard

PROGRAMMING LANGUAGE(S) USED:
PL/1
Statistical Analysis System (SAS)

PLACE OF DEVELOPMENT:
Language Processing Center
School of Languages and Linguistics
Georgetown University
Washington, D.C. 20057

PREVIOUS RELEASE DATE:
December, 1981

PROPRIETARY RIGHTS HELD BY:
Shaefer, Leonard
Language Processing Center
School of Languages and Linguistics
Georgetown University
Washington, D.C. 20057

PRIMARY ACCESSION CODE: DA005

TITLE OR NAME:
VARBRUL: A Tool for Analyzing Language Variability

PROGRAM CREATOR(S)/DEVELOPERS:
Sankhoff, G.
Sankhoff, D.
Cedergren, H.
Labov, W.

PROGRAMMING LANGUAGE(S) USED:
FORTRAN IV

PLACE OF DEVELOPMENT:
Not available

PREVIOUS RELEASE DATE:
1975

PROPRIETARY RIGHTS HELD BY:
Sankhoff, David
Centre de Recherches Mathematiques
Universite de Montreal, and
Department of Linguistics
University of Pennsylvania

PRIMARY ACCESSION CODE: DA007

TITLE OR NAME:
Vocabulary Intensity Analyzer

PROGRAM CREATOR(S)/DEVELOPER(S)
Kidder, Carole

PROGRAMMING LANGUAGE(S) USED:
PL/1

PLACE OF DEVELOPMENT:
Penn State University

PREVIOUS RELEASE DATE:
1974

PROPRIETARY RIGHTS HELD BY:
Kidder, Carole
Syntactic Density and Vocabulary
Intensity Program
Lock Haven State College
Lock Haven, PA 17745

PRIMARY ACCESSION CODE: DA008

TITLE OR NAME:
ESANALYSIS

PROGRAM CREATOR(S)/DEVELOPER(S)
Hertz, Robert

PROGRAMMING LANGUAGE(S) USED:
A.P.L.

PLACE OF DEVELOPMENT:
California State University

MOST RECENT RELEASE DATE:
1981

PROPRIETARY RIGHTS HELD BY:
Hertz, Robert
English Department
California State University
Long Beach, CA 90840

PRIMARY ACCESSION CODE: DA009

TITLE OR NAME:
CYNTHIA (This is the "mother" program for accession codes DA010 thru DA017)

PROGRAM CREATOR(S)/DEVELOPER(S)
Hendon, Rufus S.

PROGRAMMING LANGUAGE(S) USED:
IBM System/360 Assembler Language

PLACE OF DEVELOPMENT:
Yale University, Dept. of East Asian Languages and Literatures

MOST RECENT RELEASE DATE:
1981

PROPRIETARY RIGHTS HELD BY:
Hendon, Rufus S.
Yale University
Dept. of East Asian Languages
and Literatures
3963 Yale Station
New Haven, Connecticut 06520

PRIMARY ACCESSION CODE: DA010

TITLE OR NAME:
CYNKWIC

PROGRAM CREATOR(S)/DEVELOPER(S):
Hendon, Rufus S.

PROGRAMMING LANGUAGE(S) USED:
IBM System/360 Assembler Language

PLACE OF DEVELOPMENT:
Yale University, Dept. of East Asian Languages and Literatures

MOST RECENT RELEASE DATE:
1981

PROPRIETARY RIGHTS HELD BY:
Hendon, Rufus S.
Yale University
Dept. of East Asian Languages and Literatures
3963 Yale Station
New Haven, Connecticut 06520

PRIMARY ACCESSION CODE: DA011

TITLE OR NAME:
CYNMERGE

PROGRAM CREATOR(S)/DEVELOPER(S):
Hendon, Rufus S.

PROGRAMMING LANGUAGE(S) USED:
IBM System/360 Assembler Language

PLACE OF DEVELOPMENT:
Yale University, Dept. of East Asian Languages and Literatures

MOST RECENT RELEASE DATE:
1981

PROPRIETARY RIGHTS HELD BY:
Hendon, Rufus S.
Yale University
Dept. of East Asian Languages and Literatures
3963 Yale Station
New Haven, Connecticut 06520

PRIMARY ACCESSION CODE: DA012

TITLE OR NAME:
CYNFREQ

PROGRAM CREATOR(S)/DEVELOPER(S):
Hendon, Rufus S.

PROGRAMMING LANGUAGE(S) USED:
IBM System/360 Assembler Language

PLACE OF DEVELOPMENT:
Yale University, Dept. of East Asian Languages and Literatures

MOST RECENT RELEASE DATE:
1981

PROPRIETARY RIGHTS HELD BY:
Hendon, Rufus S.
Yale University
Dept. of East Asian Languages and Literatures
3963 Yale Station
New Haven, CT 06520

PRIMARY ACCESSION CODE: DA013

TITLE OR NAME:
CYNDEX

PROGRAM CREATOR(S)/DEVELOPER(S):
Hendon, Rufus S.

PROGRAMMING LANGUAGE(S) USED:
IBM System/360 Assembler Language

PLACE OF DEVELOPMENT:
Yale University, Dept. of East Asian Languages and Literatures

MOST RECENT RELEASE DATE:
1981

PROPRIETARY RIGHTS HELD BY:
Hendon, Rufus S.
Yale University
Dept. of East Asian Languages and Literatures
3963 Yale Station
New Haven, CT 06520

PRIMARY ACCESSION CODE: DA014

TITLE OR NAME:
CYNTEXT

PROGRAM CREATOR(S)/DEVELOPER(S):
Hendon, Rufus S.

PROGRAMMING LANGUAGE(S) USED:
IBM System/360 Assembler Language

PLACE OF DEVELOPMENT:
Yale University, Dept. of East Asian Languages and Literatures

MOST RECENT RELEASE DATE:
1981

PROPRIETARY RIGHTS HELD BY:
Hendon, Rufus S.
Dept. of East Asian Languages and Literatures
3963 Yale Station
New Haven, CT 06520

PRIMARY ACCESSION CODE: DA015

TITLE OR NAME:
DIACYNKW(B)

PROGRAM CREATOR(S)/DEVELOPER(S):
Hendon, Rufus S.

PROGRAMMING LANGUAGE(S) USED:
IBM System/360 Assembler Language

PLACE OF DEVELOPMENT:
Yale University, Dept. of East Asian Languages and Literatures

MOST RECENT RELEASE DATE:
1981

PROPRIETARY RIGHTS HELD BY:
Hendon, Rufus S.
Yale University
Dept. of East Asian Languages and Literatures
3963 Yale Station
New Haven, CT 06520

PRIMARY ACCESSION CODE: DA016

TITLE OR NAME:
DIACYNFQ

PROGRAM CREATOR(S)/DEVELOPER(S):
Hendon, Rufus S.

PROGRAMMING LANGUAGE(S) USED:
IBM System/360 Assembler Language

PLACE OF DEVELOPMENT:
Yale University, Dept. of East Asian Languages and Literatures

MOST RECENT RELEASE DATE:
1981

PROPRIETARY RIGHTS HELD BY:
Hendon, Rufus S.
Yale University
Dept. of East Asian Languages and Literatures
3963 Yale Station
New Haven, CT 06520

PRIMARY ACCESSION CODE: DA017

TITLE OR NAME:
ILINDEX

PROGRAM CREATOR(S)/DEVELOPER(S):
Hendon, Rufus S.

PROGRAMMING LANGUAGE(S) USED:
IBM System/360 Assembler Language

PLACE OF DEVELOPMENT:
Yale University, Dept. of East Asian Languages and Literatures

MOST RECENT RELEASE DATE:
1981

PROPRIETARY RIGHTS HELD BY:
Hendon, Rufus S.
Yale University
Dept. of East Asian Languages and Literatures
3963 Yale Station
New Haven, CT 06520

PRIMARY ACCESSION CODE: DA018

TITLE OR NAME:
RANCORK

PROGRAM CREATOR(S)/DEVELOPER(S):
Hendon, Rufus S.

PROGRAMMING LANGUAGE(S) USED:
PL/1 and IBM System/360 Assembler

PLACE OF DEVELOPMENT:
Yale University, Dept. of East Asian Languages and Literatures

MOST RECENT RELEASE DATE:
1981

PROPRIETARY RIGHTS HELD BY:
Hendon, Rufus S.
Yale University
Dept. of East Asian Languages and Literatures
3963 Yale Station
New Haven, CT 06520

PRIMARY ACCESSION CODE: DA020

TITLE OR NAME:
CMATCH

PROGRAM CREATOR(S)/DEVELOPER(S):
Kittredge, Richard

PROGRAMMING LANGUAGE(S) USED:
FORTRAN IV

PLACE OF DEVELOPMENT:
Contrastive Syntax Project
University of Montreal

MOST RECENT RELEASE DATE:

PROPRIETARY RIGHTS HELD BY:
Kittredge, Richard
Department of Linguistics
University of Montreal
C.P. G128
Montreal, H3C, 3J7
Quebec

PRIMARY ACCESSION CODE: DA021

TITLE OR NAME:
LISTER

PROGRAM CREATOR(S)/DEVELOPER(S):
Rand, Earl

PROGRAMMING LANGUAGE(S) USED:
IBM System/360 Assembler Language

PLACE OF DEVELOPMENT:
UCLA

MOST RECENT RELEASE DATE
1981

PROPRIETARY RIGHTS HELD BY:
Rand, Earl
TESL
Rolfe 3303, UCLA
Los Angeles, CA 90024

PRIMARY ACCESSION CODE: DA022

TITLE OR NAME:
INDEX

PROGRAM CREATOR(S)/DEVELOPER(S):
Rand, Earl

PROGRAMMING LANGUAGE(S) USED:
IBM System/360 Assembler Language

PLACE OF DEVELOPMENT:
UCLA

MOST RECENT RELEASE DATE:
1981

PROPRIETARY RIGHTS HELD BY:
Rand, Earl
TESL
Rolfe 3303, UCLA
Los Angeles, CA 90024

PRIMARY ACCESSION CODE: DA023

TITLE OR NAME:
COCOA

PROGRAMMING LANGUAGE(S) USED:
FORTRAN

PLACE OF DEVELOPMENT:
Atlas Computer Laboratory
Chilton, Didcot, Oxfordshire
England

MOST RECENT RELEASE DATE:
Not available

PROPRIETARY RIGHTS HELD BY:
Atlas Computer Laboratory
Chilton, Didcot
Oxfordshire, England

PRIMARY ACCESSION CODE: DA024

TITLE OR NAME:
OCP (Oxford Concordance Program)

PROGRAMMING LANGUAGE(S) USED:
FORTRAN

PLACE OF DEVELOPMENT:
Oxford University Computing
Service

MOST RECENT RELEASE DATE:
1981

PROPRIETARY RIGHTS HELD BY:
Hockey, Susan
Oxford University Computing Service
13 Banbury Rd.
Oxford, England

PRIMARY ACCESSION CODE: DA025

TITLE OR NAME:
CONCORD

PROGRAM CREATOR(S)/DEVELOPER(S):
Rand, Earl

PROGRAMMING LANGUAGE(S) USED:
WATFIV (FORTRAN IV)

PLACE OF DEVELOPMENT:
UCLA

MOST RECENT RELEASE DATE:
1981

PROPRIETARY RIGHTS HELD BY:
Rand, Earl
TESL
Rolfe 3303, UCLA
Los Angeles, CA 90024

PRIMARY ACCESSION CODE: DA026

TITLE OR NAME:
Vocabulary Analysis

PROGRAM CREATOR(S)/DEVELOPER(S):
Moe, Dr. Alden

PROGRAMMING LANGUAGE(S) USED:

PLACE OF DEVELOPMENT:
Purdue University

MOST RECENT RELEASE DATE:
Not available

PROPRIETARY RIGHTS HELD BY:
Dahlin, Gerry M.
Control Data Corporation
4201 Lexington Ave.
North Arden Hills, MN 55112

PRIMARY ACCESSION CODE: DA027

TITLE OR NAME:
Word String Analysis

PROGRAM CREATOR(S)/DEVELOPER(S):
Moe, Alden

PROGRAMMING LANGUAGE(S) USED:
Not available

PLACE OF DEVELOPMENT:
Purdue University

PREVIOUS RELEASE DATE:

FOR FURTHER INFORMATION, CONTACT:
Dahlin, Gerry M.
Control Data Corporation
4201 Lexington Ave.
North Arden Hills, MN 55112

PRIMARY ACCESSION CODE: DA028

TITLE OR NAME:
Inquirer III (Edinburgh Version)

PROGRAM CREATOR(S)/DEVELOPER(S):
Stone, P.J.

PROGRAMMING LANGUAGE(S) USED:
PL/1

PLACE OF DEVELOPMENT:
Harvard University

PREVIOUS RELEASE DATE:
1975

FOR FURTHER INFORMATION, CONTACT:
Coxon, A.P.M.
Department of Sociology
University College
Cardiff, England

PRIMARY ACCESSION CODE: DA030

TITLE OR NAME:
PARTS: A System for Assigning Word Classes to English Text

PROGRAM CREATOR(S)/DEVELOPER(S):
Cherry, L.L.

PLACE OF DEVELOPMENT:
Bell Laboratories, Murray Hill, New Jersey

PREVIOUS RELEASE DATE:
June, 1978

FOR FURTHER INFORMATION, CONTACT:
Cherry, L.L.
Bell Laboratories
Murray Hill, NJ

PRIMARY ACCESSION CODE: DA031

TITLE OR NAME:
STYLE and DICTION

PROGRAM CREATOR(S)/DEVELOPER(S):
Cherry, Lorinda L.
Vesterman, W.

PLACE OF DEVELOPMENT:
Bell Laboratories, Murray Hill, New Jersey

PREVIOUS RELEASE DATE:
February, 1981

FOR FURTHER INFORMATION, CONTACT:
Cherry, L.L.
Bell Laboratories
Murray Hill, NJ

PRIMARY ACCESSION CODE: DA032

TITLE OR NAME:
Linguistic String Parser

PROGRAM CREATOR(S)/DEVELOPER(S):
Sager, Naomi
Grishman, Ralph

PROGRAMMING LANGUAGE(S) USED:
FORTRAN

PLACE OF DEVELOPMENT:
New York University

PREVIOUS RELEASE DATE:
Not available

PROPRIETARY RIGHTS HELD BY:
Sager, Naomi, Director
New York University String Project
Warren Weaver Hall
New York, NY 10003

PRIMARY ACCESSION CODE: DA034

TITLE OR NAME:
CLAS: A Computerized Language Analysis System

PROGRAM CREATOR(S)/DEVELOPER(S):
Watts, James J.
Borden, George A.

PROGRAMMING LANGUAGE(S) USED:
PL/1

PLACE OF DEVELOPMENT:
Pennsylvania State University

PREVIOUS RELEASE DATE:
1971

FOR FURTHER INFORMATION, CONTACT:
Borden, George A.
Associate Professor of Speech
Pennsylvania State University
State College, PA

PRIMARY ACCESSION CODE: DA035

TITLE OR NAME:
SATO: A System for Text and Content Analysis

PROGRAM CREATOR(S)/DEVELOPER(S):
Meunier, Jean-Guy
Rolland, Stanislas
D'Aoust, Francois

PROGRAMMING LANGUAGE(S) USED:
FORTRAN
COMPASS (Machine Language of CDC Cyber Computers)

PLACE OF DEVELOPMENT:
Universite du Quebec, Montreal

PREVIOUS RELEASE DATE:
1973

PROPRIETARY RIGHTS HELD BY:
Meunier, Jean-Guy
Department of Philosophy
Universite du Quebec
Montreal, Canada

PRIMARY ACCESSION CODE: DA036

TITLE OR NAME:
BALCON: A Bilingual Analytical Literary and Linguistic Concordance

PROGRAM CREATOR(S)/DEVELOPER(S):
Hockey, Susan M.

PROGRAMMING LANGUAGE(S) USED:
FORTRAN

PLACE OF DEVELOPMENT:
Oxford University Computing Service, England

PREVIOUS RELEASE DATE:
1975

PROPRIETARY RIGHTS HELD BY:
Hockey, Susan M.
Oxford University Computing Service
13 Banbury Rd.
Oxford, England

PRIMARY ACCESSION CODE: DA037

TITLE OR NAME:
SELECT

SUBTITLE:
A Computer Program to Identify Associationally Rich Words For Content Analysis.

PRIMARY AUTHOR(S):
Iker, H.P.

PLACE OF DEVELOPMENT:
Not available

MOST RECENT RELEASE DATE:
1975

PRIMARY ACCESSION CODE: DA038

TITLE OR NAME:
The Automatic Discourse Analysis Method (ADA)

FOREIGN-LANGUAGE TITLE OR NAME:
L'Analyse Automatique Du Discourse (AAD)

PRIMARY AUTHOR(S):
Haroche, Claudine
Pecheux, Michel

PLACE OF DEVELOPMENT:
Centre National De La Recherche Scientifique
Paris, France

MOST RECENT RELEASE DATE:
1972

PRIMARY ACCESSION CODE: DA039

TITLE OR NAME:
STAR: A Computerized Reading Level Analysis Program

PRIMARY AUTHOR(S):
Walker, Noojin
Boillot, Michel

PROGRAMMING LANGUAGE(S) USED:
Not available

PLACE OF DEVELOPMENT:
General Motors Corporation

PREVIOUS RELEASE DATE:
1978, January

FOR FURTHER INFORMATION, CONTACT:
Walker, Noojin
Boillot, Michel
Pensacola Junior College
Pensacola, Florida

PRIMARY ACCESSION CODE: DA040

TITLE OR NAME:
MEDSIRCH: Multiple Choice Test Items

PROGRAM CREATOR(S)/DEVELOPER(S):
Hazlett, C.B.

PROGRAMMING LANGUAGE(S) USED:
FORTRAN IV

PLACE OF DEVELOPMENT:
University of Alberta,
Edmonton, Alberta

PREVIOUS RELEASE DATE:
Not available

FOR FURTHER INFORMATION, CONTACT:
Hazlett, C.B.
University of Alberta
Edmonton, Alberta
Canada

PRIMARY ACCESSION CODE: DA041

TITLE OR NAME:
MENTREX

PROGRAM CREATOR(S)/DEVELOPER(S):
Not available

PROGRAMMING LANGUAGE(S) USED:
Not available

MOST RECENT RELEASE DATE:
1973

FOR FURTHER INFORMATION, CONTACT:
Libaw, Frieda B.
Mentrex Enterprises
Los Angeles, CA

PRIMARY ACCESSION CODE: DA042

TITLE OR NAME:
Computer Program to Prepare
Cloze Texts

PROGRAM CREATOR(S)/DEVELOPER(S):
Hines, Theodore C.
Warren, Jerry

PROGRAMMING LANGUAGE(S) USED:
SNOBOL 4
SPITBOL
PL/I

MOST RECENT RELEASE DATE:
1978

FOR FURTHER INFORMATION, CONTACT:
Hines, Theodore C.
Warren, Jerry
Library of Science/Educational
Technology Division
School of Education,
University of North Carolina
Greensboro, North Carolina

PRIMARY ACCESSION CODE: DA043

FOREIGN-LANGUAGE TITLE OR NAME:
JEUDEMO

SUBTITLE:
Systeme De Traitement De Texte

PROGRAM CREATOR(S)/DEVELOPER(S):
Quellette, Francine

PROGRAMMING LANGUAGE(S) USED:
Not available

PLACE OF DEVELOPMENT:
Centre De Calcul, Universite
De Montreal, Canada

PREVIOUS RELEASE DATE:
Not available

PRIMARY ACCESSION CODE: DA044

TITLE OR NAME:
The London Text Analysis Program

SUBTITLE:

PROGRAM CREATOR(S)/DEVELOPER(S):
Rushby, N.J.

PROGRAMMING LANGUAGE(S) USED:
FORTRAN IV

PLACE OF DEVELOPMENT:
National Development Programme
In Computer Assisted Learning
London, England

PREVIOUS RELEASE DATE:
1974

PRIMARY ACCESSION CODE: DA045

TITLE OR NAME:
SPSS: The Statistical Package For The Social Sciences

PROGRAM CREATOR(S)/DEVELOPER(S):
Nie, N.
Bent, D.H.
Hull, C.H.

PLACE OF DEVELOPMENT:
The National Opinion Research Center, Chicago, Illinois

MOST RECENT RELEASE DATE:
1972

FOR FURTHER INFORMATION CONTACT:
Dahlin, Gerry M.
Control Data Corporation
4201 Lexington Ave.
North Arden Hills, MN 55112

PRIMARY ACCESSION CODE: DA046

TITLE OR NAME:
The Automated Examination Generator (AEG)

PROGRAM CREATOR(S)/DEVELOPER(S):
Ansfield, Paul J.
Rushby, N.J.

PLACE OF DEVELOPMENT:
National Development Programme in Computer Assisted Learning, London, England

MOST RECENT RELEASE DATE:
1974

FOR FURTHER INFORMATION, CONTACT:
Ansfield, Paul J.
University of Wisconsin
Oshkosh, Wisconsin

PRIMARY ACCESSION CODE: DA047

TITLE OR NAME:
WORDCOUNT

PROGRAM CREATOR(S)/DEVELOPER(S):
Hertz, Robert

PROGRAMMING LANGUAGE(S) USED:
A.P.L.

PLACE OF DEVELOPMENT:
California State University

MOST RECENT RELEASE DATE:
1981

PROPRIETARY RIGHTS HELD BY:
Hertz, Robert
English Department
California State University
Long Beach, CA 90840

PRIMARY ACCESSION CODE: DA049

TITLE OR NAME:
LEXICON

PROGRAM CREATOR(S)/DEVELOPER(S):
Hertz, Robert

PROGRAMMING LANGUAGE(S) USED:
A.P.L.
BASIC

PLACE OF DEVELOPMENT:
California State University

MOST RECENT RELEASE DATE:
1981

PROPRIETARY RIGHTS HELD BY:
Hertz, Robert
English Department
California State University
Long Beach, CA 90840

PRIMARY ACCESSION CODE: DA050

TITLE OR NAME:
TRANSFORMATIONS

PRIMARY AUTHOR(S):
Hertz, Robert

PROGRAMMING LANGUAGE(S) USED:
A.P.L.

PLACE OF DEVELOPMENT:
California State University

MOST RECENT RELEASE DATE:
1981

PROPRIETARY RIGHTS HELD BY:
Hertz, Robert
English Department
California State University
Long Beach, CA 90840

PRIMARY ACCESSION CODE: DA051

TITLE OR NAME:
SEARCH

PRIMARY AUTHOR(S):
Hertz, Robert

PROGRAMMING LANGUAGE(S) USED:
A.P.L.

PLACE OF DEVELOPMENT:
California State University

MOST RECENT RELEASE DATE:
1981

PROPRIETARY RIGHTS HELD BY:
Hertz, Robert
English Department
California State University
Long Beach, CA 90840

PRIMARY ACCESSION CODE: DA052

TITLE OR NAME:
An Automated Recognition Grammar for English

PROGRAM CREATOR(S)/DEVELOPER(S):
Culicover, P.
Kimbal, J.
Lewis, C.
Loveman, D.
Moyne, J.

PROGRAMMING LANGUAGE(S) USED:
PL/1

PLACE OF DEVELOPMENT:
IBM Corporation

PREVIOUS RELEASE DATE:
1969

FOR FURTHER INFORMATION, CONTACT:
Culicover, P.
Kimbal, J.
Lewis, C.
Loveman, D.
Moyne, J.
IBM Corporation

PRIMARY ACCESSION CODE: DA053

TITLE OR NAME:
PSIII: Protosynthex III

PROGRAM CREATOR(S)/DEVELOPER(S):
Not available

PROGRAMMING LANGUAGE(S) USED:
Not available

PLACE OF DEVELOPMENT:
Not available

PREVIOUS RELEASE DATE:
1976

FOR FURTHER INFORMATION, CONTACT:
Not available

PRIMARY ACCESSION CODE: DB001

TITLE OR NAME:
QUIZTEXT

PROGRAM CREATOR(S)/DEVELOPER(S):
Hendon, Rufus S.

PROGRAMMING LANGUAGE(S) USED:
IBM System/360 Assembler Language

PLACE OF DEVELOPMENT:
Yale University, Dept. of East Asian Languages and Literatures

MOST RECENT RELEASE DATE:
1981

PROPRIETARY RIGHTS HELD BY:
Rufus S. Hendon
Yale University
Dept. of East Asian Languages and Literatures
3963 Yale Station
New Haven, CT 06520

PRIMARY ACCESSION CODE: DB003

TITLE OR NAME:
RATSALL

PROGRAM CREATOR(S)/DEVELOPER(S):
Smith, John B.

PROGRAMMING LANGUAGE(S) USED:
PL/1

PLACE OF DEVELOPMENT:
Dept. of English, Penn State U., University Park, PA 16802

PREVIOUS RELEASE DATE:
Not available

PROPRIETARY RIGHTS HELD BY:
John B. Smith
Dept. of English, Penn. State U.
University Park, PA 16802

PRIMARY ACCESSION CODE: DB004

TITLE OR NAME:
An Interactive System for the Taxonomic Analysis of Natural Language

PRIMARY AUTHOR(S):
Ratcliff, Bryan
Yandle, John R.

PROGRAMMING LANGUAGE(S) USED:
FORTRAN

PLACE OF DEVELOPMENT:
Computer Centre, University of Birmingham

MOST RECENT RELEASE DATE:
1973

PROPRIETARY RIGHTS HELD BY:
Not available

PRIMARY ACCESSION CODE: DB005

TITLE OR NAME:
TCAP: Test Construction and Analysis Program

PROGRAM CREATOR(S)/DEVELOPER(S):
Baker, Frank B.
Martin, T.J.

PROGRAMMING LANGUAGE(S) USED:
FORTRAN

PLACE OF DEVELOPMENT:
University of Wisconsin, Madison, WI

PREVIOUS RELEASE DATE:
1972

PROPRIETARY RIGHTS HELD BY:
Baker, Frank B.
University of Wisconsin
Madison, WI

PRIMARY ACCESSION CODE: DB006

TITLE OR NAME:
ILIAD

PROGRAM CREATOR(S)/DEVELOPER(S):
Wilson, Kirk
Bates, Madeleine

PROGRAMMING LANGUAGE(S) USED:
INTERLISP
Prototype in UCSD-PASCAL

PLACE OF DEVELOPMENT:
Boston University

MOST RECENT RELEASE DATE:
Feb. 1980

FOR FURTHER INFORMATION, CONTACT:
Kirk Wilson
ILIAD Project
Boston University
1019 Commonwealth Ave.
Boston, MA 02215

*U.S. GOVERNMENT PRINTING OFFICE : 1982 0-372-091/933